National Defense Research Institute

Effec... ...rain, Man... ...actics, and ... ...SR on the Effectiveness of Long-Range Precision Fires

A Stochastic Multiresolution Model (PEM) Calibrated to High-Resolution Simulation

Paul K. Davis, James H. Bigelow, Jimmie McEver

RAND

Prepared for the
Office of the Secretary of Defense

This study was motivated by the results of high-resolution simulations of long-range precision fires that were employed against an invader marching through mixed terrain. Although effectiveness was expected to be less than for desert-terrain cases, it proved *much* less than anticipated (DSB, 1998a,b). It was clear that many factors were at work, so providing a physical explanation and projecting results for other circumstances was not straightforward.

We therefore began to develop a multiresolution family of models to better understand the phenomena, permit the broad-ranging exploratory analysis for which high-resolution simulation is inappropriate, and suggest priorities for field experiments. If successful, our work would illustrate concretely how such a family-of-models approach—coupled with experiments—could be taken routinely to improve military analysis and its underlying military science.

This report, then, describes a fast-running, stochastic, multiresolution desktop model (PEM) and its calibration to data from high-resolution simulation. We also describe a simplified and deterministic "Repro model" called RPEM for possible use in more aggregated campaign-level models such as JICM or JWARS. PEM and RPEM could substantially improve the defense community's ability to reflect, in routine analysis, many effects of C^4ISR, the maneuver tactics of the invasion force, and relatively detailed characteristics of the long-range fires. However, more empirical work and high-resolution simulations are also badly needed.

Our work was accomplished as part of a special crosscutting project sponsored by the advisory group of the National Defense Research

Institute (NDRI), which is RAND's federally funded research and development center (FFRDC) for the Office of the Secretary of Defense, Joint Staff, unified commands, and defense agencies. Comments are welcome and should be addressed to the principal author in Santa Monica, CA (e-mail: pdavis@rand.org).

CONTENTS

FIGURES

TABLES

OBJECTIVES

This report describes how various situational and tactical factors—which are usually treated only in complex models, if at all—can influence the effectiveness of long-range precision weapons in interdicting a moving armored column. The variables we consider are characteristics of the C⁴ISR system—i.e., the system for Command, Control, Communications, Computers, Intelligence, Surveillance, and Reconnaissance. The variables treated include: time from last update; missile and weapon characteristics, such as footprint; maneuver pattern of the advancing column, such as vehicle spacing, aggregate terrain features, such as open versus mixed terrain of different types; and employment tactics for long-range fires, such as firing in salvos with the missiles offset in time and space.

We also describe a stochastic personal computer model (PEM, which stands for PGM effectiveness modifier) to explore these effects systematically.[1] PEM was motivated by and has been calibrated to results of high-resolution simulation at the entity level. It is quite useful for scaling calculations, although absolute weapon-effectiveness levels also depend on classified details that we have not modeled, such as the acoustic environment, which is a function of the larger march configuration and terrain in the general area of targeting. Finally, we also describe a simplified and deterministic "Repro model"

---

[1]PEM is programmed in Analytica®, a visual-modeling system for Macintosh and PC computers, which is available from Lumina Decision Systems.

called RPEM and provide reductionist results in a set of tables. Such simplified representations of the phenomena may prove directly useful in higher level campaign models such as JICM and the emerging JWARS.

CONCLUSIONS

General

The factors mentioned above have very large implications for the effectiveness of long-range precision fires, as measured by kills per two-missile salvo or per aircraft sortie. If a "standard" case for such fires is attacking a column of armored vehicles separated on average by no more than 50 meters and traveling across terrain that offers no concealment (e.g., the desert), then effectiveness can be quite high—either with small-footprint air-delivered munitions such as sensor-fused weapons (SFWs) or with long-range missiles such as ATACMS/BAT. However, under other assumptions regarding stand-off range, C⁴ISR capabilities, dispersal, and terrain, effectiveness can drop by two orders of magnitude.

The various factors interact in complex ways that cannot be modeled as a mere product of, for example, a terrain adjustment, a Red dispersion adjustment, a C⁴ISR adjustment, and so on. The sensitivity of outcome to one factor depends strongly on other factors, so linear sensitivity analysis around some baseline can be misleading. For this and other reasons related to input uncertainty, we have emphasized an "exploratory analysis approach" that assesses relative effectiveness across a wide range of cases in which the factors are varied simultaneously.

Specific Observations

Dispersal During Maneuver. The Red maneuver pattern, which can reflect a passive Red countermeasure against the Blue attack, interacts with several of the other factors. If Blue's timing of weapon delivery is good, changing the Red maneuver pattern—e.g., by increasing the spacing between armored fighting vehicles (AFVs)—usually has a much greater effect in canopied terrain with small open areas than in terrain with large ones. Moreover, in such terrain, a weapon such as ATACMS/BAT with a large footprint loses much of its advantage over a weapon, such as sensor-fused weapons, with a small footprint.

C⁴ISR Factors. Depending on the other factors, the time since last update[2] can range from very important to irrelevant. If Red maneuvers in long columns of densely spaced vehicles, then kills per salvo or sortie are independent of the time since last update. But if Red has a more complex maneuver pattern, Blue must time his shots so that they arrive in open areas when the Red packet does. Accurate timing of shots becomes even more important if open areas are small, at least in the sense that kills per shot decline by a larger fraction as time since last update increases. But if open areas are small enough, kills per shot may be too meager for shooting to be worthwhile with weapons intended for multiple kills, even if the time since last update is zero. This is especially true if the weapon has a long "descent time"—i.e. a long time between when its submunitions acquire targets and when they reach the ground.

If the Red formation maintains strict discipline and moves with constant speed, then a small time since last update will pay large dividends for "normal" maneuver tactics and open terrain. But if Red is more dispersed and deliberately changes speeds frequently, effectiveness of area weapons in mixed terrain can be quite low even if time from last update is rather small. In that case the payoff is much higher for less expensive one-on-one weapons, which may be delivered from short range by aircraft or from the short-range fire of maneuver forces.

The ability to discriminate between live and dead targets is significant if multiple shots are fired into the same area without leaving time for dead targets to stop and cool, or if a single weapon can kill a large number of vehicles. It is an open issue, however, whether adding such discrimination capabilities will prove cost-effective.

---

[2]The time of last update is the time between when a weapon is last directed to impact at a particular time and place and when impact occurs. It is an important attribute of the C⁴ISR-weapon-system combination. It can be shortened by: minimizing command-related delays, time of flight, and processing within the C⁴ISR system; providing target updates to missiles in flight; and combinations.

Force-Employment Tactics. Offsetting a salvo's missiles in time usually has only a marginal effect by reducing the likelihood of a second missile attacking a portion of a packet or packet group that has already been depleted. Offsetting small-footprint area weapons delivered from aircraft is quite important because dead-target effects (failure to discriminate) would otherwise greatly reduce effectiveness.

Summary Quantitative Results

Tables S.1 and S.2 summarize results of PEM runs in estimating the effectiveness of ATACMS/BAT and F-16/SFW weapon combinations versus the attacker's choices of AFV spacing, the type of terrain, and the time from last update for the interdictor's weapon system. Since actual effectiveness numbers would also depend on both classified and unclassified details not reported here (e.g., the acoustic environment due to the particular types of vehicles in the march, their configuration, and their interaction with the environment), what matters most is the relative numbers within Tables S.1 and S.2. As can be seen by comparing the top-left and bottom-right figures, we should expect a factor of roughly 100 in weapon effectiveness as a function of these three variables. Comparable tables can be generated by PEM for other weapon types.

One important caution here is that readers should not attempt simple cost-effectiveness comparisons using PEM alone—even with classified input data rather than the illustrative figures we have used. As reported in an ongoing RAND study on interdiction for the Joint Staff (Ochmanek et al., unpublished), analysis strongly argues for *mixes* of different weapon types because if the United States has such mixes, and if the weapons are all of high quality, a would-be invader will have much less incentive to disperse (see also McEver, Davis, and Bigelow, forthcoming). Further, as discussed elsewhere (Matsumura, Steeb, et al., 1999), analysis also suggests that mixes of long-range fires and light-mechanized maneuver forces would have major advantages over long-range fires alone, especially in complex terrain and when the invader employs anticipated tactical and technical countermeasures.

Table S.1

Sensitivity of Kills per ATACMS/BAT Salvo to Timing Errors, Dispersion, and Type Terrain[a,b,c]

Dispersal/Terrain	Open	Mixed	Primitive Mixed
No Timing Error			
Very tight	12	10	1
Dispersed	11	6.0	0.2
Very dispersed	6.2	2.9	0.06
10 Minute Errors			
Very tight	12	10	1
Dispersed	9.1	2.2	0.3
Very dispersed	4.9	1.8	0.15
16 Minute Errors			
Very tight	12	10	1
Dispersed	6.0	3.4	0.23
Very dispersed	3.1	1.1	0.15

[a]Absolute values also depend on other situational details not provided here to avoid classification.

[b]Definitions: Very tight: 50 meters per AFV, 100 AFVs per packet; Dispersed: 100 meters per AFV, 10 AFVs per packet; Very dispersed: 200 meters per AFV, 5 AFVs per packet. Open: 12 km open-area mean widths; Mixed: 3 km open-area mean widths; Primitive: 1 km open-area mean widths. "Mixed terrain" also assumes canopies.

[c]The timing error is the difference in minutes between when the targeted packet is centered in the open area and the time of arrival of the weapon. If the error in estimating the packet's movement rate along the road is 25% (after accounting for winding roads, random movements, and deliberate changes of speed as a countermeasure), then the timing errors shown would be one-fourth the "time from last update."

Methodological Conclusions

This study demonstrates concretely the feasibility and power of a multiresolution approach to analysis that actively works the gamut from high-resolution, entity-level, man-in-the-loop simulation on the one extreme, to exploratory analysis with fast-running desktop

Table S.2

Sensitivity of Kills per F-16 Sortie with 4 SFWs to Timing Errors, Dispersion, and Type Terrain

Dispersal/Terrain	Open	Mixed	Primitive Mixed
No Timing Error			
Very tight	8.5	8.4	4
Dispersed	6.1	6.0	2.7
Very dispersed	2.7	2.6	1.0
10 Minute Errors			
Very tight	8.5	8.4	4.0
Dispersed	2	2.1	1.2
Very dispersed	0.95	0.95	0.65
16 Minute Errors			
Very tight	8.5	8.4	4
Dispersed	0.9	0.86	0.54
Very dispersed	0.36	0.36	0.25

NOTES: See Table S.1.

models on the other.[3] One of our principal objectives in undertaking this work was to accomplish such a demonstration. When such analytic work is used to inform and exploit empirical work, including large-scale field experiments, a great deal can be learned about the phenomenology of future military operations—including the risks associated with them and how to mitigate those risks. As discussed elsewhere (Davis, Bigelow, and McEver, 1999), we strongly recommend that the Department of Defense and U.S. Joint Forces Command adopt such an approach in its work on next-generation forces, doctrine, and related joint field experiments.

---

[3]For prior discussions, see Davis, Gompert, Hillestad, and Johnson (1998) and Davis, Bigelow, and McEver (1999). For underlying theory, see Davis and Bigelow (1998).

ACKNOWLEDGMENTS

We acknowledge discussions with and assistance from Randy Steeb, John Matsumura, Ernst Isensee, Tom Herbert, Scott Eisenhard, and Gail Halverson. Professors Reiner Huber (University of the Bundeswehr) and Thomas Lucas (Naval Postgraduate School) provided helpful reviews.

ACKNOWLEDGMENTS

ACRONYMS

AFV — Armored Fighting Vehicle

APC/T — Armored Personnel Carrier/Transport

ATACMS — Army Tactical Missile System

BAT — Brilliant Anti-armor Submunition

C2 — Command and Control

C^4ISR — Command, Control, Communications, Computers, Intelligence, Surveillance, and Reconnaissance

CAGIS — Cartographic Analysis Geographic Information System

EXHALT — Exploratory analysis of the Halt problem

FOFA — Follow-on forces attack

IDA — Institute for Defense Analyses

JDAM — Joint Direct Attack Munition

JICM — Joint Integrated Contingency Model

JSOW — Joint Standoff Weapon

JSTARS — Joint Surveillance and Target Attack Radar System

JWARS — Joint Warfare System

LOC — Line of Communication

MADAM	Model to Assess Damage to Armor with Munitions
MRM	Multiresolution Modeling
N/TACMS	Naval Tactical Missile System
PEM	PGM effectiveness multiplier
PGM	Precision guided munition
RMA	Revolution in military affairs
RPEM	Repro model
RSTA	Reconnaissance, Surveillance, Targeting and Acquisition
SEAD	Suppression of enemy air defenses
SFW	Sensor-fused weapon
START	Simplified Tool for Analysis of Regional Threats
UAVs	Unmanned aerial vehicles
WMD	Weapons of mass destruction

Chapter One

INTRODUCTION

OBJECTIVES

The primary purpose of this report is to describe analytically how the effectiveness of long-range precision weapons should be expected to vary when they are used against a moving armored column, depending on variables usually treated—if at all—only in much more complex simulation models. The variables we consider are:

- Characteristics of the C⁴ISR (Command, Control, Communications, Computers, Intelligence, Surveillance and Reconnaissance) system related to (1) projected target locations versus time, (2) targeting updates to en route weapons or delivery platform, (3) various delay times, and (4) likelihood of detecting and attacking a given "packet" of armored vehicles.

- *Missile/weapon characteristics* such as single-missile or single aircraft-sortie footprints, lethality against visible targets within their footprints, flight times, descent time of the weapon after final commitment to targets, accuracy, shots per salvo, and the ability to discriminate between live and dead targets.

- *Maneuver pattern* of the advancing armored column, which involves vehicle spacing, packet size (e.g., platoon size), packet configuration, packet separations, and movement rate.

- *Terrain features,* notably the length of open areas into which the missiles are targeted.

- *Tactics* involving salvo offsets.

We also have a more general methodological objective. Our work is a prototype demonstration of how high-resolution simulation can be mined for information that can then be used—albeit with caution— in fast and flexible lower-resolution depictions useful for exploratory analysis. As discussed further in reports by the National Research Council (NRC) and Defense Science Board (DSB), much more work of this sort is desirable because a deplorable gulf currently exists between studies done at the different levels of resolution. Sometimes this leads to very different perceptions of reality and disputes that should be resolvable by analysis. What is needed is an emphasis on developing multiresolution models and families of models, so that consistency can be achieved across levels of resolution and different perspectives.[1]

APPROACH

We describe the issues with a new stochastic, multiresolution model called PEM (for PGM effectiveness modifier), which is based on a simplified depiction of the problem's physics and tactics. The model is illustrated for the Joint Standoff Weapon (JSOW) using sensor-fused weapons (SFWs) and large missiles typified by the Army's Tactical Missile System with Brilliant Anti-armor Submunitions (ATACMS/BAT). Key inputs are provided parametrically to avoid classification.

PEM is calibrated from higher resolution work. The assumptions it uses for ATACMS/BAT are informed by and calibrated to results of entity-level simulation using a RAND federate of models and man-in-the-loop gaming that includes Janus, MADAM, CAGIS, and rather detailed representation of weapon and submunition characteristics (see Appendix A).[2] The assumptions for sensor-fused weapons used on aircraft-delivered weapons such as the JSOW or the Joint Direct Attack Munition (JDAM) are informed by Air Force field tests and the previous analysis by colleague Glen Kent.[3]

---

[1]Davis and Bigelow (1998), National Research Council (1997), and Defense Science Board (1998a,b).

[2]Defense Science Board (1998a,b) and Matsumura, Steeb, et al. (1999).

[3]Ochmanek, Harshberger, Thaler, and Kent (1998).

PEM provides physical insights about what can limit or enhance the effectiveness of precision fires. Quantitatively, it agrees rather well with the sparse "data" from high-resolution simulation. PEM (or a reprogrammed version)[4] could be used as a module in larger halt-phase models such as EXHALT (McEver, Davis, and Bigelow, forthcoming) or in campaign models such as JICM or Joint Warfare System (JWARS) that deal with the halt problem. Or, as shown in a later section, an even simpler "Repro model" (RPEM) motivated by results using PEM could be used in or to help calibrate such other models.

In what follows, then, we first describe the background motivating the work (Chapter Two), describe the conceptual model underlying PEM (Chapter Three), and sketch the actual PEM program (Chapter Four). Chapter Five provides details on how we analyzed high-resolution data and used them to inform and calibrate PEM. Chapter Six applies PEM and compares results to high-resolution simulation. Chapter Seven takes a final step, presenting a simplified RPEM that does a reasonably good job in representing the phenomena treated by PEM in most cases. RPEM may be used as a subroutine within other models such as EXHALT, JICM, or JWARS. PEM itself could be such a subroutine, but the reprogramming required would be more extensive. Finally, Chapter Eight summarizes the conclusions.

---

[4]PEM is programmed in Analytica®, a visual-programming system with powerful features for array mathematics and treatment of uncertainty. PEM could be reprogrammed into Visual Basic, C++, or other general-purpose languages if necessary. We understand from Roy Evans and Hank Neimeier of MITRE, who have used Analytica extensively, that it can also be reprogrammed readily into EXTEND®, which greatly decreases run time and reduces memory requirements (Belldina, Neimeier, Pullen, and Tepel, 1997). For our work, however, no such reprogramming was necessary and the advantages of the Analytica environment proved quite attractive. In the future, the content of PEM could be transferred to models such as RAND's JICM or even to the emerging JWARS.

Chapter Two

BACKGROUND

Recent years have seen considerable enthusiasm for the use of long-range precision weapons—not merely for attacking fixed elements of military infrastructure, but also for attacking invading armored columns. It has become possible to envision limiting cases in which such long-range fires—whether in the form of weapons launched from aircraft or missiles launched by Army or Navy units—might halt or severely disrupt an invading army. Even if such fires were able merely to cause substantial attrition, that might be sufficient to permit relatively small, high-quality ground forces to complete and even reverse the halt.[1]

Such long-range fires are a major element in descriptions of the so-called revolution in military affairs (RMA).[2] They have played a dominant role in many studies of the last five years, studies suggesting that small defensive forces could have extraordinary effectiveness against classic armored invasions, assuming, of course, that the defenders were suitably armed and had the necessary command, control, communications, computers, intelligence, surveillance, and reconnaissance—the unwieldy combination of functions usually referred to as C4ISR. The C4ISR might be provided by aircraft, unmanned aerospace vehicles (UAVs), satellites, or people on the ground, such as small teams of ground forces.

---

[1]See Bowie, Frostic, Lewis, Lund, Ochmanek, and Propper (1993).

[2]See, e.g., Defense Science Board (1996), Sovereign (1995), Joint Chiefs of Staff (1997), Barnett (1996), Ochmanek et al. (1998), Bingham (1997). DoD's Deep Attack Weapons Mix Study (DAWMS) reported a 1996–1997 classified multivolume study addressing many of the issues.

Various studies suggest that in the relatively near future, it should be possible for fighter aircraft to kill 1–10 armored vehicles per sortie, and for long-range missiles such as the ATACMS or a longer-range/smaller-payload Navy version dubbed NTACMS to achieve perhaps 1–10 kills per missile. Simple arithmetic would then suggest, for example, that 150 aircraft flying 2 sorties per day and killing 2 Armored Fighting Vehicles (AFVs) per sortie, plus 100 ATACMS shots per day killing 6 AFVs per shot, could kill 1,200 AFVs per day. If one assumed 600 AFVs per division and that killing half of a division's AFVs would halt it, then an 8 division attack could be stopped in 2 days! Although the numbers are contrived and real wars do not take such a simple form, the arithmetic is nonetheless instructive. Further, although large and complex simulations used by the Defense Department and its contractors may describe a much more multifaceted campaign, if they include aircraft and missiles with these kinds of capabilities, such fires can dominate the results from deep inside the simulations.

Roughly speaking, defense with such high-effectiveness precision fires succeeds quickly in simulations if the shooters and C⁴ISR are available on or shortly after D-Day. Otherwise, there can be a race and the attacker may reach his objectives, or at least engage badly outnumbered ground forces, before the fires can become sufficiently effective. Or the attacker might enter and occupy cities, from which it might be costly to evict him. An exception exists if the defender has great depth and the attacker has only one or two lines of communication (LOCs), in which case it is possible under some assumptions to use a strategy for the employment of long-range fires that could slow and even roll back the enemy's advance (Ochmanek et al., 1998; Davis, Bigelow, and McEver, 1999).

Discussions of such matters have usually focused on a replay of Desert Storm in which Iraq's forces press onward to Saudi Arabia, the Gulf Coast, and the primary oil facilities. Such scenarios play to the best case for long-range fires: long armored columns moving over long distances in the desert. Other studies have dealt with the Korean Peninsula, in which terrain and tough defensive forces would likely channelize attacking armored forces and provide opportunities for precision fires. In contrast, except for a study used for the work of this report, there has been little recent discussion about what might be expected from precision fires in "mixed terrain" of the sort that is so ubiquitous throughout the world.[3] The phrase "mixed terrain" is necessarily ambiguous because it covers many variations, but one might think here of northern Virginia (outside the urban area) or major portions of Europe. From an aerial view, one would see large expanses of wooded territory interrupted by fields, roads, and small towns. The terrain itself might have hills but would not be truly mountainous. Implicitly, at least, advocates of long-range fires have often believed that such fires would be nearly as effective in mixed terrain as in the desert.

In work supporting the 1998 summer study of the Defense Science Board, some of our RAND colleagues conducted high-resolution simulations investigating the potential effectiveness of long-range fires in mixed terrain.[4] The results were surprising and discouraging to enthusiasts of such fires: Effectiveness was down by an order of magnitude relative to earlier studies that used the same suite of models but for desert circumstances.[5] Although it was possible to provide a qualitative discussion of why the results were so bad,[6] it was clear that many effects were at work simultaneously. High-resolution simulations are simple in some respects but notoriously complex in others. Further, these simulations are not tools for quick and agile "what if?" studies. They depend on a great deal of data and prior preparation. Thus, it was obviously desirable to have a simpler model that could capture the essence of the phenomena and allow us to scale results from detailed high-resolution work to many other circumstances. It was not clear a priori, however, what could be accomplished in this regard.

In this report we describe such a model. Unlike many relatively aggregated models, this one was strongly motivated by insights gleaned

---

[3]A number of studies were done during the 1980s for the context of Europe's Central Region. At that time, however, the targets envisioned were typically very large, dense armored formations moving by Soviet doctrinal norms of the era without countermeasures against the imagined PGMs. Further, the characteristics ascribed to advanced C[4]ISR and weapon systems were more speculative. Such studies often referred to "follow-on forces attack" (FOFA) or "assault-breaker" capability.

[4]See Matsumura, Steeb, Isensee, Herbert, Eisenhard and Gordon (1999), which will also appear in DSB (1998b).

[5]Matsumura, Steeb et al. (1997), and DSB (1996, Volume 2).

[6]See DSB (1998b) or Davis, Bigelow, and McEver (1999).

from the high-resolution work. Also, we have used the high-resolution work to calibrate the principal uncertain parameters of the model. We used abstractions of some of the data used in the high-resolution work to set other parameters. Nonetheless, the model we have developed has a stand-alone conceptual basis: It is not merely some statistical "fit" but rather a depiction of the principal phenomena that appear to be at work—including various nonlinear "edge effects," stochastic features, and transitions between domains in which different variables dominate. It is likely that more extensive analysis of high-resolution data will allow us to improve the model further and sharpen its calibration, but even this initial version of PEM is quite useful.

Chapter Three

THE CONCEPTUAL MODEL

PREFACING COMMENTS

As discussed in Chapter Two, high-resolution experiments in 1998 motivated us to think more deeply about how terrain interacted with march configuration and other factors to determine the effectiveness of long-range fires. We could infer some of the issues from the initial data, but more extensive analysis would clearly be difficult, time-consuming, and ambiguous. We therefore decided to develop a theory from the physics of the problem and initial impressions from the high-resolution work, simplifying as seemed appropriate to build a tractable model (PEM) suitable for desktop calculations on a personal computer. This theory would then form the initial hypothesis against which to structure and examine the simulation data in more depth. If the model proved reasonable, we would calibrate it to the high-resolution data; if it proved erroneous, we would iterate. As it happens, the structure for PEM that emerged from theory held up reasonably well, but interpretation of variables and assignment of parameter values proved more difficult than anticipated. Further, we identified additional degrees of freedom that should be dealt with explicitly in future versions of PEM. We discuss those matters in Chapter Five. In this chapter we describe the physical picture or conceptual model underlying what emerged as PEM in its current form (PEM 1.0).

STOCHASTIC CONSIDERATIONS

PEM is a stochastic model. In principle, all of its variables can be treated as random variables characterized by probability distributions. For example, the speed with which a given unit moves through terrain is a random variable: sometimes greater and sometimes less than its mean value. The mean value itself may also be quite uncertain, especially when we deal with hypothetical future wars against hypothetical opponents. We may vary that mean value parametrically, or we may also describe its variation with a probability distribution.

This said, as we describe PEM we usually write as though the variables are deterministic so as to simplify discussion. Also, in using PEM we prefer to treat many variables deterministically to simplify analysis and reduce run times. Which variables should be treated probabilistically depends on the application.

PHYSICAL PICTURE

We consider a single avenue of approach through mixed terrain. The road is mostly canopied (at least from the perspective of a C⁴ISR system such as Joint Surveillance and Target Attack Radar System (JSTARS) viewing the road from a long standoff distance). However, it has open areas of length W within which detections can be made or attacks prosecuted (Figures 3.1 and 3.2).

The attacker configures his columns to have platoons, companies, and higher level units. Each platoon has N_{tot} vehicles, of which N are AFVs with the remainder being, e.g., trucks and jeeps. We focus on the AFVs. Each platoon or "packet" has a length L dictated by the number of vehicles and the average spacing S. The entire column moves, during a maneuver period, at some average speed V_{ave} except when moving through open areas within which packets are vulnerable to attack. A given packet moves through an open area with a speed V. The attacker may shorten his packets, increasing the density of AFVs. This allows him to push more force through the area quickly but may increase vulnerability. The attacker may also mix non-AFVs with AFVs within individual packets, thereby diluting the nominal value of packets, but at some unknown price to movement efficiency (Figure 3.2).

RAND *MR1138-3.1*

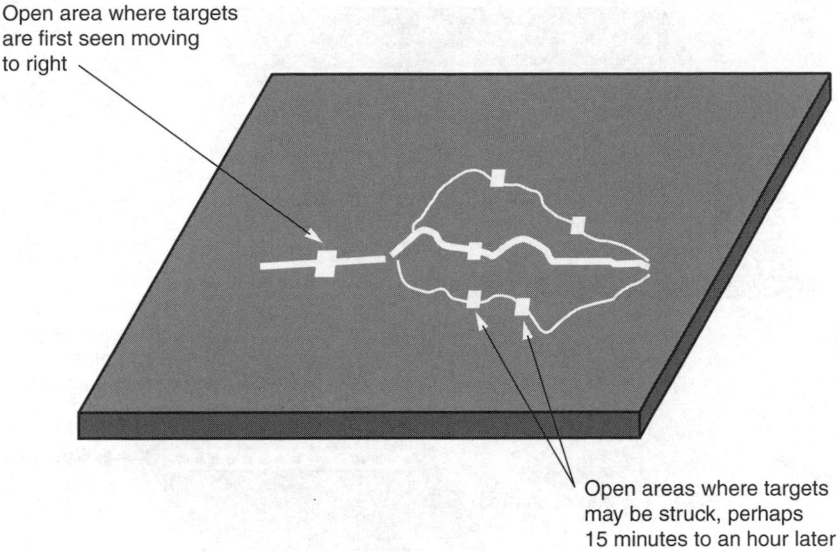

Figure 3.1—Targeting Vehicles Moving in Canopied Terrain with
Occasional Open Areas

TIMING THE ATTACK

The C[4]ISR system detects a given packet and estimates its time of arrival at an open area farther down the avenue of approach. A missile or salvo of missiles is fired, arriving at that open area at time T after the final estimate. T can be the sum of command-control delays and flight time. Or, if the missile can be given updated targeting information while en route and adjust its impact time and point, T is the time between the last update and impact.

The missile is aimed to land at the center of the open area at precisely the time when the packet in question will also be centered there (Figure 3.3). However, the missile's impact point is subject to error (Figure 3.4) due to imperfect accuracy, and its time of impact is subject to error because of the difficulty of estimating—perhaps 5–30 minutes in advance—when the packet will in fact be in the middle of the area. The error in estimated impact time (TOA_error) grows with

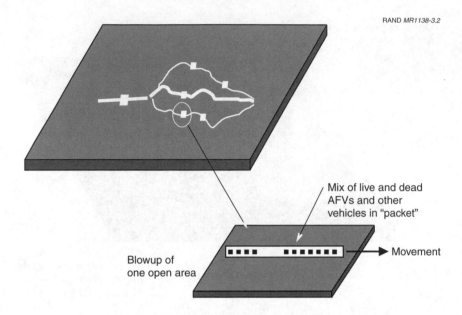

RAND *MR1138-3.2*

Figure 3.2—Configuration at Time of Weapon Impact

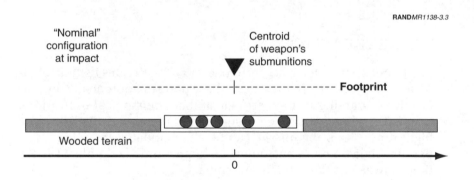

RAND*MR1138-3.3*

Figure 3.3—Ideal Geometry of Impact

T, the fractional error in estimating the effective speed of the packet and the packet's actual speed. The missile has a footprint dimension along the axis of the road that is F in diameter. The missile is effective only if it is able to detect and track targets for a period $T_{descent}$ before impact. In Figure 3.4, this excludes portions

RANDMR1138-3.4

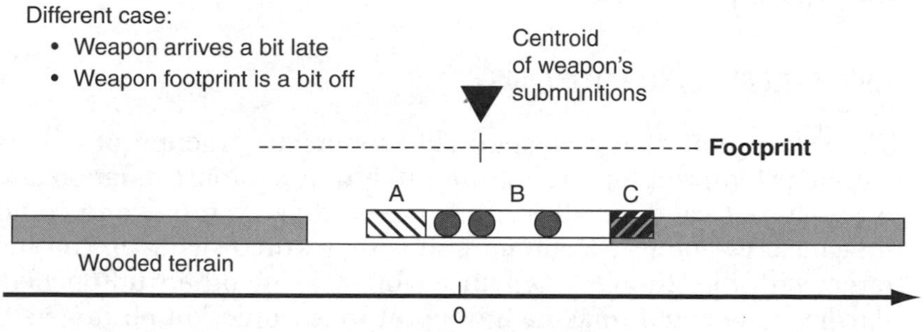
Different case:
• Weapon arrives a bit late
• Weapon footprint is a bit off
Centroid of weapon's submunitions
Footprint
A
B
C
Wooded terrain
0

Figure 3.4—Impact Geometry for Imperfect Targeting and Missile Accuracy

A and C of the packet: Part C has entered the woods,[1] whereas part A was in the woods at time $-T_{descent}$ (Figure 3.5).

When the missiles' weapons have impact, we assume that the submunitions will kill some constant fraction FracKill of the armored vehicles within the weapon's footprint F (up to a maximum based on the number of submunitions and other factors) but only if the

RANDMR1138-3.5

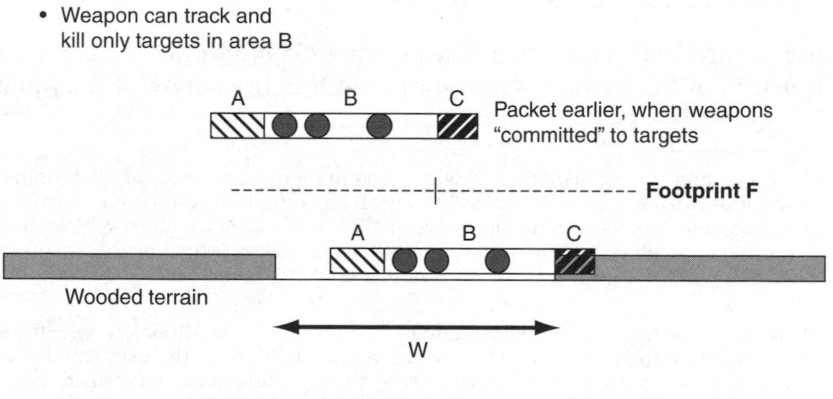
• Weapon can track and kill only targets in area B
A
B
C
Packet earlier, when weapons "committed" to targets
Footprint F
A
B
C
Wooded terrain
W

Figure 3.5—Effect of Finite Weapon "Descent Time"

---

[1]Although our illustrations all refer to wooded terrain, the same arguments apply for urban terrain.

targets are all in the open and, again, have been in the open for a time $T_{descent}$ (Figure 3.5).[2]

LETHALITY ASSUMPTIONS

The assumption that a weapon kills a constant fraction of targets within its footprint (up to a maximum) is only an approximation and is surely not valid for all weapon types. For example, one could imagine a weapon with submunitions that would "gang up" on the target with the strongest signature while leaving others untouched. Further, one could imagine a weapon that would kill no targets if only a very small number were present. Such a weapon might not have a good enough signal-to-noise ratio to start the process of orienting its submunitions. PEM users should consider fine-tuning the FracKill variable based on high-resolution simulation data for the specific munition of interest.[3] The function FracKill(N), where N is the number of AFVs in the killing zone, would be directly determined from the high-resolution data. In this report we have assumed constant FracKill, subject to a maximum determined by the number of submunitions, missile and bus reliability, and so on. We have then parameterized these inputs to avoid potential classification.[4]

ESTIMATING THE "KILLING ZONE"

In this physical picture, then, the effective "killing zone" is a complex function of the weapon's footprint length F, the length of the open

---

[2]This time must be estimated based on a much more detailed study of the terminal behavior of weapons, such as is possible with high-resolution simulation. It can be short (e.g., ten seconds) or rather long (e.g., a minute or so). Since target vehicles may be moving at roughly 1 km per minute, and since open areas sometimes are relatively short with lengths on the order of 1–4 km, this nonzero descent time can be significant.

[3]This would be appropriate for a study to determine the conditions to which the particular munition is well suited. For other kinds of studies, the user could vary FracKill to find regions of model space where it makes a difference. This would be appropriate to match munition characteristics to terrain type and Red formations.

[4]As discussed in Chapter Five and Appendix C, FracKill turns out to depend strongly, for some weapons such as BAT, on factors that we cannot yet model in the aggregate, notably the acoustic environment, which depends on the general background of noise determined by the overall movement and, e.g., presence of trees and urban structures.

area W, the weapon's descent time $T_{descent}$, the targets' speed as they cross the open area V (which may or may not be the same as their average movement rate $V_{maneuver}$ between targeting and weapon impact), and the actual point and time of impact relative to the ideal—the center of the open area upon which the target packet is aimed. The impact point and time are, of course, random variables dependent on weapon accuracy and the ability to predict target movement rates. In Figure 3.5, the killing zone corresponds to the portion of the packet tagged B.

Which factor is limiting in determining the length of the killing zone depends, then, on many factors. For example, in the desert, W would be quite large and would not be limiting. In road marches in which units follow each other closely, the length of a packet would have no meaning and would not be limiting.[5] The size of a weapon's footprint will typically be important, but how much footprint is enough depends on the other factors such as the length of open areas.

Figure 3.6 summarizes the dynamics schematically.

DIMINISHING RETURNS FOR SALVOS

If the missiles are fired in salvos—intended to arrive at the same time and place—there will be diminishing returns from one missile to the next because each subsequent missile sees a smaller density of targets within the killing zone. Precisely how much smaller the density will be depends not only on weapon effectiveness but also on salvo geometry. Figure 3.7 sketches such a geometry with the second missile's killing zone being part B of the higher rectangle and the first missile's killing zone being part B of the lower one. (Note that this figure, unlike Figure 3.5, shows the locations of the packet at the two impact times, not the impact time and time $-T_{descent}$ for a single missile.) As before, the noses and tails (parts A and C in this figure) are not in the killing zone because of having entered the wooded terrain by the time of impact or having been in the wooded terrain at

---

[5]Recent work by colleagues David Ochmanek, Glenn Kent, and others refer to BAT as a one-on-one weapon, which is understandable for their applications to desert terrain and relatively open mountain roads. In that limiting case, kills per BAT could be constant for a wide range of packet size and spacing, because of BAT's large footprint and finite number of munitions.

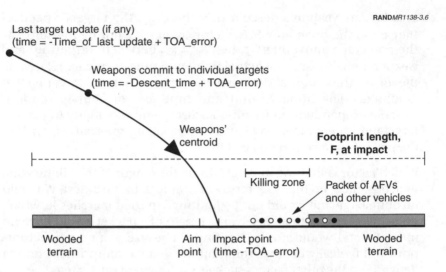

Figure 3.6—Summary Depiction of Dynamics

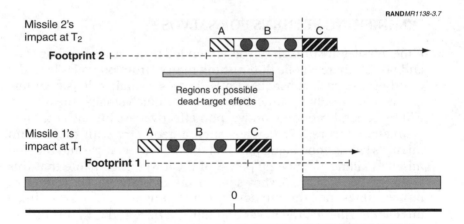

Figure 3.7—Geometry of a Two-Weapon Missile Salvo

time $-T_{descent}$. Some of the packet visible to the first weapon may enter foliage before the second weapon has impact, but other parts of the packet may have been invisible to the first but are visible to the second.[6]

Another factor in the salvo calculation is that some weapons may attack targets already disabled or in the process of stopping. This dead-target effect appears to be the largest cause of diminishing returns, unless the weapons are able to discriminate between live and dead targets. Such discrimination depends not only on differences in the behavior and characteristics of live versus dead vehicles, but also on the kind of sensor one uses to detect those differences. For example, a live moving target will be hot, and hence detectable by an infrared sensor. The same sensor will overlook a cold dead target. But it may mistake a burning dead vehicle for a live one, and will overlook a live target that has been stopped long enough to cool down. Good discrimination is likely to require multisensor suites. The cost-effectiveness of improved discrimination by the weapons is unclear.

To estimate the dead-target effect, we can assume that weapons are equally likely to attack live or dead targets within a given group of packets. We also assume that a given open area is not repeatedly targeted within a half hour or so because, in that case, we would expect substantially degraded results (as have been observed in high-resolution simulation). This assumption should be revisited if discrimination capabilities improve or if firing successively at particular open areas appears to be important for lack of other options.[7]

The norm is two missiles per salvo for weapons like ATACMS.[8] Results for a salvo can sometimes be improved by offsetting the in-

---

[6]This effect has not been included in the baseline version of PEM because it did not seem to be sufficiently large to bother, and because it substantially complicates the calculations when—as discussed later—we include the effects of hitting adjacent packets.

[7]A more detailed treatment could parameterize the relative likelihood of munitions hitting live or dead targets as a function of how long dead targets have been dead. We did not see much value in such detail in our work.

[8]This doctrinal norm was developed years ago when the targets in mind were dense formations of Soviet armored forces. It seemed reasonable to require a high damage expectancy against a targeted cluster of vehicles, which led to the two-missile-per-

tended position and/or time of impact. This can improve the odds that at least one of the two missiles has targets in its killing zone, that the second missile will not see a depleted portion of the packet, and that the second missile's killing zone is not cluttered with "dead" targets. Precisely what can be accomplished here, however, depends on many weapon details and the physical configuration. In an early version of PEM, we calibrated FracKill (the fraction of killing-zone targets killed by a single missile) by dividing high-resolution results for a two-missile salvo by two. For the baseline PEM model, however, we have changed this assumption, drawing from limited high-resolution experiments on the issue to estimate a better FracKill for a single missile. The baseline model then calculates an estimated per-salvo value that reflects a dead-target effect reducing the second missile's effectiveness. Multi-missile salvos are an expensive way to use weapons like ATACMS when the target array is sparse. Although discriminating between live and dead targets may be possible technologically, we are skeptical about glib claims on the matter.[9]

In representing the use of air-delivered sensor-fused weapons (SFWs), we drew upon the published work of Glenn Kent (Ochmanek et al., 1998). Here we assumed—merely for the sake of a baseline—that a "nominal" weapon delivery would be four SFWs (an F-16's payload) delivered well but imperfectly along 0.4 km of a packet's line of advance. Based on a combination of Air Force weapon-range tests and calculations, such weapon employment should provide "availability kills" of about 70 percent of the armored vehicles in that length of column.

---

salvo standard. The norm should be rethought for attacks on more highly dispersed formations.

[9]Sometimes, those promising future discrimination have in mind using moving-target-indicator (MTI) radars to distinguish between live targets that are still moving and dead targets that have stopped. That should be possible, with sufficient investment, but it would invite the countermeasure of stopping temporarily while under attack. Discrimination is sometimes postulated to depend on detecting smoke or fire from dead vehicles. However, many "killed vehicles" (especially those with mere mobility kills) may emit neither smoke nor fire: Indeed, they may seem to be in good shape visually except for one or more small holes. This said, dead targets that have been in place for some time will be cooler and may be easily avoided.

HITTING PACKETS OTHER THAN THOSE TARGETED

If packets move in groups (e.g., the platoons of a company), it is possible for a shot intended to hit one group to instead strike the previous one or the subsequent one. In principle, the shot might hit a much earlier or much later packet. For dispersed formations, the likelihood of this becomes one of random success rather than something more systematic. In PEM we consider only the kills achieved within a given grouping of three packets (e.g., a company of three platoons). We show results for hitting the packet targeted and results for the case in which there are packets immediately ahead and immediately behind the targeted packet (i.e., with a packet-to-packet separation of perhaps 1–4 kilometers), with bigger separations at the next level of organization. If the first packet of a group is attacked, then there will be no prior packet in the same group. Thus, this "other-packet" adjustment may overestimate the effect of hitting adjacent packets. Viewing results with and without the "other-packet" adjustment, however, should bound the calculations.[10]

OTHER FEATURES OF PEM

This completes the description of the core elements of the model. In addition, PEM provides some rough estimates (based on limited high-resolution work) of how the likelihood of shooting at a given packet depends on the qualitative level of C^4ISR and how the overall attrition of a force moving through the battle zone depends also on how many opportunities there are to shoot at a given packet (the number of open areas along the avenue). We do not elaborate on those features here.

CAVEAT

Models change and documentation can often not keep up. Thus, readers interested in using PEM should consult the actual PEM program as the definitive source of information. Analytica's self-

---

[10]The exception here is for near-continuous, high-density movements. They can be treated simply by setting the number of AFVs per packet to a large number.

documenting features make study of the program itself much easier than with traditional programming languages.

Chapter Four

THE PEM MODEL AS PROGRAMMED

Having described the conceptual model underlying PEM, let us now describe briefly its implementation as a computer program in the Analytica programming system. This documentation is not complete because part of the philosophy underlying the use of Analytica is that the program itself is understandable and substantially self-documenting. Analytica achieves this via its dependence on influence diagrams and a number of built-in features, such as automatic listing of each variable's inputs and outputs. All of the higher level information needed to understand Analytica qualitatively is included here.[1]

DATA DICTIONARY

Table 4.1 is an unofficial data dictionary. It is unofficial in the sense that it was generated for the manuscript, whereas the official data dictionary could be generated automatically from the program itself at any given time.[2] The first column of Table 4.1 shows the "identifier" of each datum, variable, and module (each "node" in the terminology of Analytica). The second column gives the node's math symbol, which is used in this report in preference to computer-style names. The third column is the node's title as it appears on the

---

[1]Analytica programs can be read and operated with a free "viewer" version of Analytica that can be downloaded from the developer's web site at www.lumina.com.

[2]To generate the rigorous data dictionary for PEM at any given time, we use a simple PERL script developed by colleague Manuel Carrillo to supplement Analytica. We include the script with copies of PEM.

Table 4.1

Unofficial Data Dictionary

Identifier	Math Symbol	Title, Description of Variable	Comment
Enemy Maneuver Tactics			
Perspec_of_packets		Perspective of packet calculation	A switch determining which packet variables are inputs and which are calculated
AFV_spacing	S	Spacing between AFVs (km)	
Afvs_pp	N	AFVs per packet	Can be specified rather than calculated
Afvs_pp_input	N	AFVs per packet (input)	Can be specified rather than calculated
Fraction_afvs	F_{afvs}	Fraction of packet's vehicles that are AFVs	
Speed_maneuver	$V_{maneuver}$	The maneuver force's average speed from the time of targeting until time of weapon impact	Drops out of calculations if time from last update and fractional prediction error are used.
Speed_across	V	Speed of packet as it moves across open area (km/minute)	May or may not be same as average speed
Vehicles_per_packet		Vehicles per packet (= N/F_{afvs})	Includes AFVs, trucks, jeeps, non-armored command vehicles
Packet_length_input	L	Length of packet (km) (= $(N + 1/2)S$)	Can be specified rather than calculated
Closure_tactics		Closure tactics	After suffering attrition, the attacker can either close ranks, reconstructing whole packets, or move with "holes" in packets.

Table 4.1 (continued)

Identifier	Math Symbol	Title, Description of Variable	Comment
		C4ISR Factors	
Res_of_last_update		Resolution of last-update calculation	A switch determining resolution. Time of last update can be calculated or specified
Time_of_update_list		List of last-update times (minutes)	If time of last update is specified, then this is the list of parameter values available
Frac_predict_error	E	Fractional error in estimating packet's speed from last update until centered in target area	Is straightforward only when packet is moving in straight line at constant speed. Terrain, orientation of sensor line of sight, and tactics can confuse issue
Enroute_targeting		Enroute targeting?	A resolution switch; if yes, then the time of last enroute update is specified
Time_of_last_enroute	T	Time of last enroute update if there is one (minutes)	
Extra_c2_time		Decision time (minutes) before firing above and beyond time for processing of RSTA data	Can be many minutes in current procedures; could be nearly zero
Latency_of_rsta_data		Latency of RSTA data (minutes after data was valid)	Time for RSTA and weapon systems to process targeting data and prepare firing

Table 4.1 (continued)

Identifier	Math Symbol	Title, Description of Variable	Comment
Packet_detec_prob		Packet detection probability (rough placeholder for what should be better calibrated input from high-resolution simulation) (a normal distribution)	Probability density for a packet moving through an open area being detected and tracked; if it is, track is assumed to be maintained
Missile/Weapon Inputs			
Res_impact_time		Resolution of impact-time calculation	A switch dictating whether impact time will be calculated or specified directly
Flight_time		Flight time (minutes)	Flight time of missile from launch until impact
Descent_time	$T_{descent}$	Descent time (time from commitment of submunitions until impact)	
Weapon_accuracy		Weapon accuracy (km)	One-sigma error around intended impact point (the center of the target area), assuming normal distribution
Footprint_length	F	F (km)	Length of footprint along road. Targets outside footprint will definitely not be attacked
Frac_kill_input	Z	Fraction of targets killed within footprint	Can be specified parametrically or as a list

Table 4.1 (continued)

Identifier	Math Symbol	Title, Description of Variable	Comment
Source_of_frackill		Source of Frac_kill	Switch dictating whether one uses the value calibrated to high-resolution results of the 1998 study or specifies this parametrically
Terrain-Related Factors			
Length_of_open_area	W	Length of open area (km)	Could be given as a distribution for a known region, assuming tactics
Salvos_per_packet		Number of successive salvos as packet traverses multiple open areas (no more than one salvo per target area-packet)	Considered a function of "environment," but could also be function of firing doctrine (wastefulness)
Res_pkt_tgting		Resolution of packet targeting calculation	Resolution switch; fraction of packets engaged can be an input, or it can be calculated
Mean_areas_per_av		Mean number of target areas per avenue of approach	Used in log-normal distribution for number of target areas through which a given packet moves
Gstev_for_areas		The geometric standard deviation for the number of target areas per avenue of approach	

diagrams. The title is an alias for the identifier. The fourth column is a terse description. Fuller descriptions are found within the program itself.

DATA FLOW DIAGRAMS

Figure 4.1 is the top-level view of PEM. Figure 4.2 shows the contents of the module called interface. This shows most but not all of the inputs and a few of the model's outputs. The interface module is a convenient place to review and change model assumptions. The entire module can be saved and stored independently of the model, which is convenient for maintaining configuration control and a library of cases.

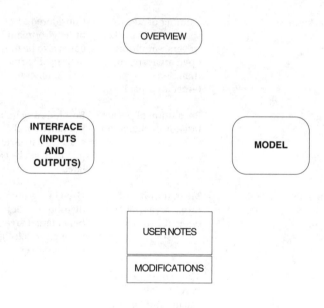

Figure 4.1—Top-Level View of PEM

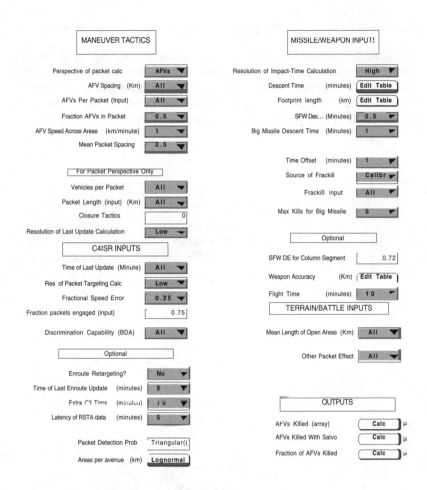

Figure 4.2—User Interface of PEM

The Overview, Modifications, and User Notes modules of Figure 4.1
contain documentation. The model itself is contained within
"Model." To provide an overview of its structure, we show data flow
diagrams in what follows. They play a central role in software engi-
neering, along with identifying objects and certain other features of
programs. When programming in Analytica, these diagrams are part
of the model itself, not something generated separately.

If we "double-click" on the module "Model" we obtain Figure 4.3.

As Figure 4.3 indicates, the model proceeds by calculating the arrival time and impact point of weapons, relative to when the targeted packet is at the center of the open area (feedbacks not shown). PEM also calculates the packet size. These and other factors determine the AFVs killed by a shot or salvo of shots as shown in Figure 4.4.

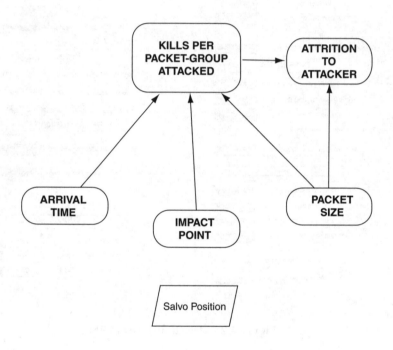

Figure 4.3—Top-Level Data Flow of PEM

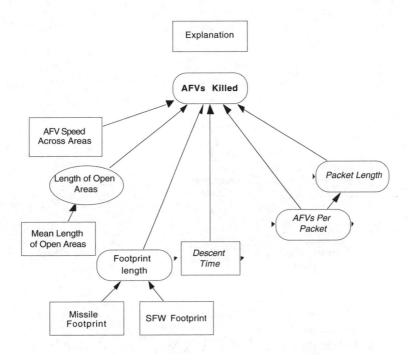

Figure 4.4—Kills per Packet Group Attacked

Returning to Figure 4.3, let us now elaborate on the nature of the three modules Arrival Time, Impact Point, and Packet Size. Figure 4.5 shows the content of Arrival Time. Note the multiresolution design allowing the user to input variables as detailed as time of flight (bottom right) or, instead, to input higher level variables such as Time of Last Update, which otherwise would be calculated from below. By inputting higher level variables, one greatly reduces the number of degrees of freedom, thus simplifying explanation and exploratory analysis (Davis and Bigelow, 1998).

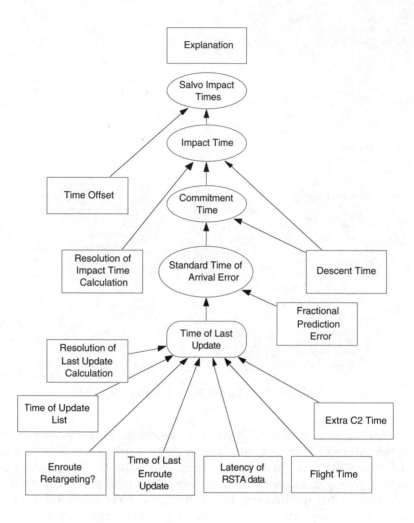

Figure 4.5—Data Flow Within Arrival Time Module

As Figure 4.6 indicates, impact points are simply a function of the weapons' accuracy. There may be a bias error (mean impact point other than zero). Beyond that, it is assumed that impact points are normally distributed with a standard error.

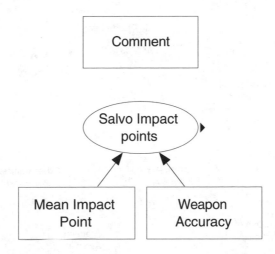

Figure 4.6—Data Flow Within Impact Point Module

Figure 4.7 shows that Packet Length may be calculated or inputted directly. There are alternative perspectives allowed for here. Which combination of inputs one uses depends on "perspective." Ordinarily we suggest using AFV spacing, AFVs per packet, and fraction AFVs in packet. This uniquely determines packet length and other vehicles per packet.

Figure 4.8 shows how PEM calculates an adjustment factor for the number of AFVs killed by the second missile in a salvo of two. Currently, we assume that a pro rata share of the second missile's submunitions will be wasted on vehicles already killed by the first missile. Or, more precisely, we assume that kills are proportional to the fraction of AFV targets in the killing zone that are live.

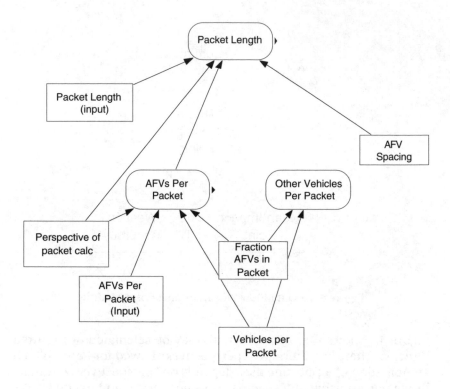

Figure 4.7—Data Flow Within Packet Size Module

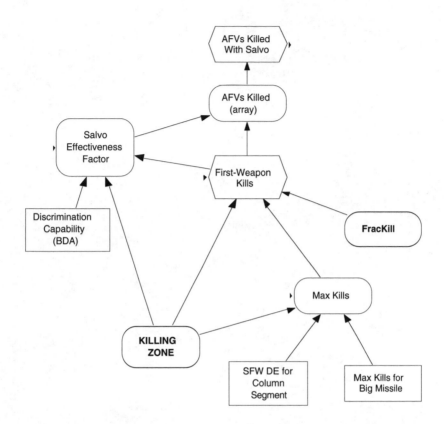

Figure 4.8—AFVs Killed by Single Weapon and by Salvo

Chapter Five

CALIBRATION FROM HIGH-RESOLUTION
SIMULATION

INTRODUCTION

Previous chapters have described the PEM model. In this chapter we discuss our detailed analysis of "data" generated from high-resolution simulations accomplished for the 1998 DSB summer study (Mastumura et al., 1998). We approached that analysis with a preliminary version of PEM in mind. That helped structure the analysis by identifying questions that we should address. For example, PEM postulated that the AFV kills achieved by a single large missile such as ATACMS with a BAT warhead would be proportional to the number of AFVs in the killing zone (up to a maximum). That might or might not have been consistent with simulation data. If PEM's structural assumptions proved reasonable, we would proceed to calibrate its key parameters. If, instead, we found serious inconsistencies, we would then revise our thinking, adjust the structure of PEM, and iterate.

For much of the work, we treated the simulation data as though they were experimental data from the real world—without examining details of the simulation's algorithms (although we discussed such issues qualitatively with RAND's Janus team). The simulation suite (Appendix A) has been used in many prior studies for the Army, Air Force, DARPA, and the Joint Staff and has gained considerable credibility as the result of numerous reviews conducted within those studies—often with the participation of individuals with strong motivation to find errors because of the potential implications for weapon-system trade-offs.

In some cases, however, initial analysis raised perplexing questions, and we made special high-resolution runs to clarify either phenomenology or the workings of the simulation. This work was accomplished by Gail Halverson with the oversight assistance of Tom Herbert.

As it happened, the principal features of PEM held up reasonably well, but many details proved much more complicated than we initially expected. Furthermore, some of the intuitive concepts represented in PEM turned out to be "fuzzy." That is, it was difficult to define precisely what some of PEM's variables correspond to in the physical world. This is common in cookie-cutter models in which weapons are effective within a certain radius and ineffective outside of it, but more subtle problems arose related to the scale of observation.

In what follows we summarize our data-analysis experience. We have omitted a discussion of our many false starts and our difficulties mapping the data structures of the simulation into our needs. We have attempted to convey, however, a sense of how we compared hypotheses against data, and how we chose to make simplifying interpretations. Further, we have indicated numerous instances in which real-world empirical information is needed because even current-day high-resolution simulation is not reliable.

PRINCIPAL VARIABLES OF ANALYSIS

Our data analysis was largely organized around an effort to validate the appropriateness and calibrate the values of eight of PEM's parameters, notably:

W = Length of open areas (meters)

N = Number of AFVs per packet

S = Separation of AFVs within a packet (meters)

P = Separation of packets, tail to head (meters)

V =   Speed of packet crossing an open area (meters/sec)

T =   Time since last update (sec)[1]

F =   Diameter of weapon's footprint (meters)

FracKill   =   Fraction of AFVs in the effective footprint that are killed by that weapon (here the "effective footprint" is the "killing zone" of the PEM model)

Of these, only the last was what one might ordinarily think of as a simulation output. The others seemed more like inputs. However, quantities such as the separation between AFVs was in fact an output: The simulation started with an initial march configuration, but considerable randomization occurred as the simulation proceeded and, for example, distances between units in a packet and distances between packets changed as the result of random processes and interactions with terrain (e.g., slowing during uphill road movements). One objective of data analysis, then, was to determine what PEM's inputs should be to represent the situation described by the simulation.

LENGTH OF OPEN AREAS

The DSB '98 high-resolution simulations represented an area 200 km x 200 km with four million cells, each of which corresponded to a 100 x 100 meter square of ground. Each cell was coded in one of three ways: Tree, City, or Open. If a vehicle found itself in a cell coded Tree or City, it was invisible and invulnerable to the long-range weapons.

---

[1]The "real" variable here is error in estimating when the weapon should arrive. This depends on the product of the packet's maneuver speed, the interdictor's fractional error in estimating that speed, and the distance between targeting and weapon impact. The fractional speed error, in turn, depends not only on characteristics of the radar system being used and geometry, but also on detailed knowledge of the road and on whether the packet's maneuver speed is constant for the time of interest. The timing error is also equal to the product of the time from last update (measured backward from time of impact) and the fractional error in estimating speed. Because the time from last update is a meaningful parameter in discussing C4ISR systems, we have chosen to focus on it and hold constant (at 0.25) an estimate of fractional error based on discussions with the officer who did the man-in-the-loop targeting in the high-resolution work. We note that other studies may assume a much smaller fractional error, perhaps because of focusing strictly on radar capabilities.

If it was in a cell coded Open, the weapons could detect and kill it with some probability. Of the four million cells in the 200 x 200 km box, 61 percent were coded Open, 38 percent Tree, and 1 percent City. The city-coded cells may have played a more important role than the 1 percent might suggest, however, because cities tend to be located in relatively large open areas (i.e., without tree cover) and roads are routed through cities. Thus cities occupy many of the areas that would otherwise be preferred targets.

Overlaid on this terrain was a network of roads. In these particular simulations, vehicles were assumed to travel on the roads; they did not travel cross-country. Of course, different routes across the 200 x 200 km box would contain different amounts of open area, so by selecting his routes, Red had some control over the fraction of time each vehicle was vulnerable. In the DSB '98 simulations, the Red force split into three columns, following roughly parallel routes from Southwest to Northeast (marked as bold in Figure 5.1).

Blue could choose which open areas to fire at. If Blue never fired at open areas shorter than 2,000 meters, then it might seem that the length of the open areas to be used in PEM should be at least 2,000 meters—even if many open areas were smaller (we discuss caveats below). Note that in such simulations, Blue's and Red's tactics both affected what PEM's input should be in this regard: Red could choose corridors with smaller or larger open areas, making trade-offs with highway width and quality. Blue could choose what threshold of open-area length to use in deciding whether to attempt targeting vehicles. Thus, one cannot determine the open-area length to be used in an analysis from merely looking at maps.

Looking along the routes, one finds that defining an open area was not straightforward. Figure 5.2 shows a small area along the Southern route, at a scale that reveals all the detail of the high-resolution simulation's terrain representation.

We first asked, is the portion of the route in the lower left of the figure, which follows a narrow gap in the trees, an "open area"? From the point of view of the man-in-the-loop who decided when and where to fire TACMS, this was *not* an open area. He worked from maps at a much larger scale, on which gaps like this one were too

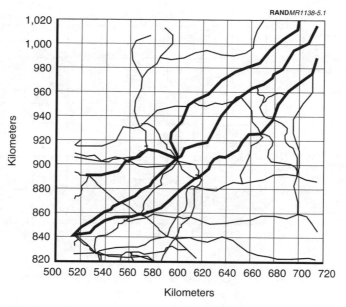

(Numbers along axes are positions, in Km, from a reference point.)

Figure 5.1—Road Network with Routes Followed by Three Red Columns

narrow to show. So none of Blue's TACMS salvos were aimed deliberately at narrow gaps.[2]

This said, the picture could be quite different from the point of view of the TACMS missile and its BAT submunitions. If a real-world missile arrived over the section shown, and some of the gap was within the missile's footprint, would vehicles in the gap be vulnerable? This was beyond the ability of the present simulation to answer, and it was beyond the scope of our study to investigate the matter.

---

[2]This was probably realistic for a variety of reasons, including the fact that normal highways (as distinct from superhighways) paralleled by tall trees would not be consistently visible from a standoff airborne radar 100 miles or so away.

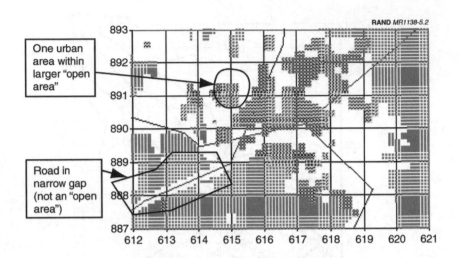

Figure 5.2—Detail of an Open-Area Portion of a March Route

Second, and more relevant to our study, we asked whether the part of the route that passes in and out of city-coded cells consists of one open area or several. On the map used by the man-in-the-loop, the urban area a bit above and to the left of the middle of Figure 5.2 did not even show up, so this area appears to be a single open stretch of road five kilometers long.[3] In contrast, to a TACMS missile it may appear to be several open areas with widths from 200 meters to about a kilometer. Perhaps two open areas separated by a sufficiently short stretch of city look like a single open area to BAT, but in this particular case, that seems unlikely. And perhaps BAT cannot even "see" a sufficiently short open area.

More work is needed to understand such issues for any given area of interest. Moreover, empirical work in the field is needed to assure that even high-resolution simulations are correct.

Based on our data analysis, if we calibrated PEM to agree with what the man-in-the-loop was seeing, we might use a uniform distribution

_____

[3]This is due to the fact that the icon that represented a city-occupied cell did not contrast well with the color chosen to depict an open cell. Choosing a different color scheme could make cities more visible, but problems of resolution would remain.

between 3.33 km and 6.66 km. However, we have tentatively con-cluded that the effective open-area lengths suitable to PEM should be smaller for most of the open areas because of the urban structures within many of what appeared to the man-in-the-loop to be the best targets.

Summary: For the cases for which we had high-resolution data, we characterized W in PEM by using either (a) parameterized open-area lengths of 1, 3, 5, and 7 km, or (b) a triangular probability distribution with a mode of 1, 3, or 5 km, and with minimum and maximum values at 0 km and the mode plus 2 km, respectively.

RED'S MARCH DOCTRINE AND THE VARIABLES N, S, P, AND V

Appendix B describes in detail how we calibrated these four parameters in PEM from the results of the high-resolution simulations for the DSB '98 study. In principle, we probably could have estimated them from input data reported for the simulations. However, operating in the spirit of experimental data analysis, we examined the simulation output directly. As it turned out, this proved quite useful and changed our impressions of what had gone on in the simulation: In this case, as in many high-resolution simulations, behavior of entities was more complex than one expects from examining input assumptions.

In the DSB '98 simulations, there were 543 Red vehicles of 13 different types. Only 104 Red vehicles of 4 types were AFVs. As described earlier, the Red force split into three columns, following the routes shown in Figure 5.1. Each column consisted of three kinds of packets (groups of vehicles in the same platoon or section), plus some miscellaneous vehicles that did not appear to be grouped into packets. The lead packets consisted of light combat vehicles that formed a reconnaissance screen for the Red force. Next came packets of AFVs, followed by packets consisting primarily of trucks. For PEM we were interested only in the AFV packets, though in the high-resolution simulations any kind of packet could be (and was) targeted.

Packets of different kinds traveled at different speeds: reconnaissance packets at 94 km/hr, AFV packets at 76 km/hr, and truck packets at 58 km/hr. Thus, the order of the packets changed over time. Early in the simulation the chain of reconnaissance packets over-lapped the chain of AFV packets, which in turn overlapped the chain of truck packets. Later, the reconnaissance packets all pulled ahead of the lead AFV packet, and the last AFV packet passed the lead truck packet. Thus, the column, which began as a mixture of all the vehicle types, sorted itself into three more nearly homogeneous sections.[4]

As currently configured PEM represents P (the spacing between AFV packets) as a random variable, but the other parameters are deter-ministic and varied parametrically. In the DSB '98 simulations, packets were not quite so simple. First, packets of AFVs contained a variable number of vehicles (3–10, with a mean of 6.7). The spacing between AFVs was also variable (150–600 meters, with a mean of 350 meters). However, the speed of AFV packets in the DSB simulations was nearly constant (76 km/hr).

The other simplification PEM makes is to assume that all AFVs are identical. In the high-resolution simulations, there were four differ-ent kinds of AFVs. As we discuss in Appendix B, BAT detected and killed them with quite different probabilities, but PEM's aggregation appears not to be a serious problem.

Summary: In our baseline PEM for the DSB '98 cases, we either varied the number of AFVs per packet parametrically at 4, 7, and 10, or used a uniform distribution between 3 and 10 (mean value of 6.5 km). We varied the spacing of AFVs within a packet parametrically at 200, 350, and 500 meters, and for some excursion cases also considered a spac-ing of 50 meters. Packet separation was also varied parametrically at 1, 2.5, and 4 km. We used a speed of 75 km/hr (1.25 km/minute).

TIME SINCE LAST UPDATE

In the DSB '98 simulations, shots were selected by a man-in-the-loop. He would be handed a map showing the positions of Red vehicles at a simulated time t and asked to choose aim point and im-pact times for salvos of BAT-carrying TACMS missiles. For each

---

[4]We question whether faster packets could in fact pass slower ones traveling along the same road. To do so on a major highway is one thing. To do so on a country road with forest on both sides is another.

scenario there was a nominal time from last update for TACMS salvos, representing the time needed to obtain the map from the reconnaissance, surveillance, targeting, and acquisition (RSTA) assets that gathered the information, plus the delay for analysis and decisionmaking, plus the time to implement the decision, plus the flyout time for the missile. The minimum nominal delay ranged from 11 minutes to 20 minutes, depending on the case being run. The actual delay sometimes differed from the nominal delay by a minute or two, but the difference has no significant effect.[5]

Summary: By comparing PEM with DSB '98 data, we parameterized time from last update to be between 10 and 20 minutes. In looking at data for specific groups of simulations, we set them deterministically to match simulation inputs.

THE FOOTPRINT OF TACMS WITH BAT

To determine the "output" footprint of TACMS (as distinct from the complex set of inputs that represent the BAT munition and its acoustic and infra-red sensors), we calculated the distance from the aim point of each salvo in the DSB '98 simulations to the position of each Red vehicle killed by that salvo.[6] Figure 5.3 shows the results. Ninety-eight percent of kills occurred within 7.5 kilometers of the aim point. But kills were much more likely to occur within 4 to 4.5 kilometers of the aim point than farther out.

We also conducted some experiments with the high-resolution model in which we offset the aim point from a solitary Red vehicle by as much as 6.5 kilometers and still killed the vehicle. To some extent this depended on how loud the vehicle was (BAT first detects targets acoustically). Other experiments showed that when enough background noise was present, the footprint shrank, in the sense that BAT could no longer pick up a target group of vehicles from as far away.

---

[5]As the description suggests, the high-resolution work could cover only a limited number of cases and was complicated by human factors. The PEM model is quite complementary in this regard.

[6]It was occasionally uncertain which salvo was responsible for which kill, as the simulation model does not output this datum. We inferred it from the correspondence between impact time of the salvo and the kill time of the vehicle.

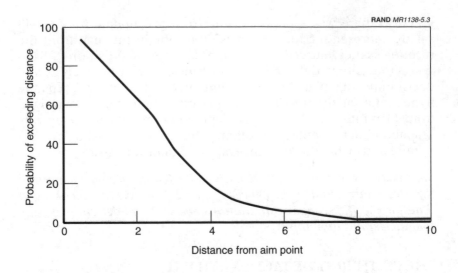

Figure 5.3—Distance Distribution from Aim Point to Kills

In PEM the footprint of a weapon is represented as a cookie cutter. There is a radius within which a vehicle has the same probability of being killed, regardless of where within that radius the vehicle is located. A vehicle outside that radius has zero chance of being killed.

Summary: When comparing PEM with DSB '98 results, we configured PEM to have a footprint diameter F of 8 km, but real-world uncertainties are large, so we varied F parametrically with values of 4, 8, and 12.

AFV FRACTION KILLED

The high-resolution model simulated in considerable detail the process by which BAT submunitions search acoustically and then home in on target vehicles by infrared sensing. Each submunition and each Red vehicle was simulated individually. In PEM, by contrast, a missile salvo kills a fixed fraction (FracKill) of the Red vehicles in the footprint, up to a maximum (MaxKills). Appendix C discusses how we calibrated FracKill and MaxKills to results from the high-resolution simulation.

Summary: Based on the DSB '98 high-resolution simulation results and cursory information about results of earlier studies (DSB, 1996), we estimated FracKill to be about 0.22 for a single missile and the fraction of AFVs killed by a salvo of two missiles to be about 0.41 for the conditions of the 1998 study. We represented FracKill for a salvo of two missiles as a stochastic variable with a triangular distribution having minimum, mode, and maximum values of 0, 0.41, and 0.82. Based on independent experiments with the high-resolution simulation, we estimated MaxKills to be about 6 AFVs for a single missile and about twice that for a salvo of two missiles. As noted earlier, this figure depended on the acoustic environment used, which we do not report here to avoid classification issues.

Chapter Six

ILLUSTRATIVE RESULTS AND DISPLAYS

This chapter presents selected results from the current version of the PEM model. First we show some comparisons of PEM with the high-resolution simulation. While the two models appear to be consistent, there are too few high-resolution results available for comparison to determine for sure over what parameter ranges PEM agrees with the high-resolution simulation.

Second, we explore the behavior of PEM over a wide range of parameter values. Through this exploratory analysis, we can develop a clear idea of which factors most strongly influence how well BAT performs. The ability to perform an exploratory analysis requiring hundreds or thousands of model runs was one of the major reasons for building an aggregate model such as PEM.

COMPARISONS OF PEM WITH HIGH-RESOLUTION SIMULATION

We select two results from the high-resolution simulation to compare with PEM. They are (1) the number of AFVs in the effective footprint of a salvo, and (2) the number of AFVs killed per salvo. Of course, we will not compare results for any particular salvo. Rather, we will compare the distribution of each of these quantities over salvos.

Figures 6.1 and 6.2 show the results from the high-resolution simulations performed for the DSB '98 summer study. In these figures we have taken the footprint to have a radius of 4 km. The most notable feature in each figure is the high probability of a "zero" result. There

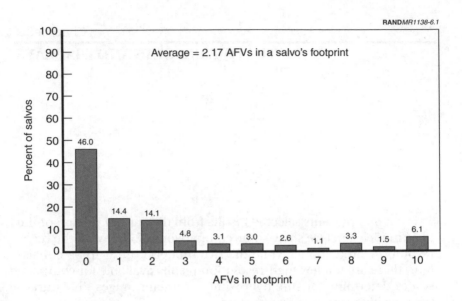

Figure 6.1—Distribution over Salvos of AFVs in Footprint

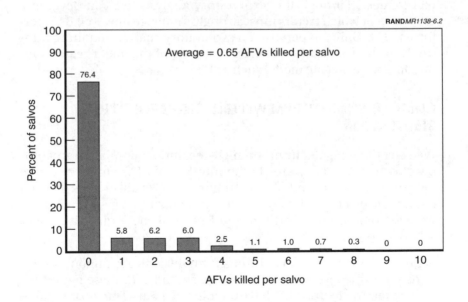

Figure 6.2—Distribution over Salvos of AFVs Killed per Salvo

were zero AFVs in the footprints of almost half the salvos and more than three-fourths of the salvos killed zero AFVs.

To compare PEM with these results, we must first decide which input values correspond to the DSB '98 simulations. It is evident from the discussion in Chapter Five that PEM will need different inputs to match different salvos from the DSB '98 cases. We ran PEM for packet spacings of 1, 2.5, and 4 km, AFV separations within a packet of 200, 350, and 500 meters, and AFVs per packet of 4, 7, and 10.

In the DSB '98 simulations, it was difficult even to establish criteria for determining what were the open areas. Measuring open stretches of road on a relatively low-resolution map indicated that the average open area was about 5 km long, though many salvos were aimed at shorter areas. But from Figure 5.2 we see that a high-resolution map might show these areas to consist of multiple smaller open stretches. We ran PEM for modal lengths of open areas of 1, 3, and 5 km.[1]

The other inputs were held constant for these runs. With the maximum open area only 7 kilometers long (see footnote), it was not necessary to consider a range of missile footprint sizes, so we set the footprint at a diameter of 8 km (radius of 4 km). We set the speed of AFV packets at a constant 75 km/hr. Since results for this type of terrain are not very sensitive to the "time since last update" parameter, unless it is made very short (a few minutes), we set it at 15 minutes.

Figures 6.3 and 6.4 show the comparisons. The bars in each figure are repeated from Figures 6.1 and 6.2; they show the results from the high-resolution simulations. Plotted over them are three distributions from PEM, one for each of the values listed above for the length of an open area. Each is formed as a weighted average of 27 PEM cases, in which packet spacing, AFV separation within a packet, and number of AFVs per packet are each varied over the three values listed earlier. We singled out the length of the open area to plot as the x axis because (1) it was the most problematic of the four parameters we varied, and (2) it seemed to make the most difference

_____

[1]In the current version of PEM, the length of open areas is a random variable with a triangular distribution. The user inputs the mode W. The minimum length is always zero and the maximum is W+2.

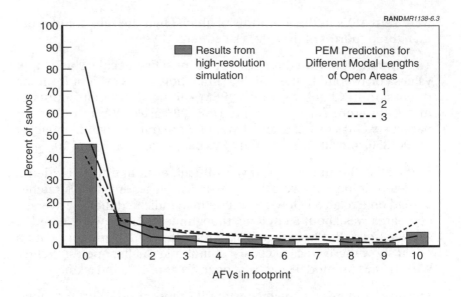

Figure 6.3—AFVs in Footprint Comparison of PEM with High-Resolution Simulation

in the distribution of both AFVs in the footprint and AFVs killed per salvo.

Clearly, PEM shares with the high-resolution simulation the characteristic that a substantial fraction of salvos have zero AFVs in their footprints,[2] and an even larger fraction of salvos kill zero AFVs. Moreover, the PEM results could be made to agree fairly closely with the high-resolution results by forming an average of the three curves in each of Figures 6.3 and 6.4 with the appropriate weights. Thus, we conclude that so far as these two quantities are concerned, PEM behaves quite similarly to the high-resolution simulation.

However, we should not expect PEM to match these results too closely. In the DSB '98 cases, there were 13 Red vehicle types in three

---

[2]In Chapter Seven we will approximate the distribution of vehicles in the footprint by a geometric distribution, whose single parameter is selected to produce a desired mean number of vehicles in the footprint. Figure 6.3 shows that this approximation introduces no great distortion.

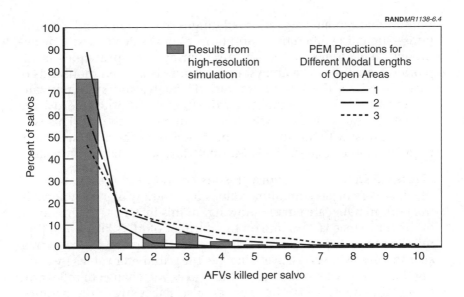

Figure 6.4—AFVs Killed per Salvo Comparison of PEM with High-Resolution Simulation

types of packets. Because in most cases the man-in-the-loop could not distinguish AFVs from other types of vehicles, some salvos were aimed at other kinds of packets. This would inflate the proportion of salvos that had zero AFVs in their footprints, which in turn would increase the proportion of salvos that killed zero AFVs (they may have killed other vehicles, however).

EXPLORATORY ANALYSIS WITH PEM

A major advantage of PEM is that one can use it to conduct exploratory analysis. Both PEM and the high-resolution simulation model it approximates are weakly predictive, at best. By this we mean that while the outcomes predicted by the model are plausible, nobody should believe the actual numerical outputs from a single case or a small number of cases to be a reliable prediction of what would happen in detail in a real-life conflict. Any defensible use of these models requires one to generalize and abstract from the actual numerical results they produce to find principles (or at least rules of thumb) that remain true for broad ranges of scenarios and assumptions—including different assumptions about scale factors. One approach that often succeeds in discovering such principles is exploratory analysis. The analyst runs hundreds or even thousands of cases in which the inputs are varied widely, and systematically searches the dataset generated from them. In other words, the model generates the data, which must then be analyzed. This is in stark contrast with the view that the model does the analysis. But it requires a model that is, like PEM, small, fast, and analytically agile.

Figure 6.5 shows an Analytica "results" display. In the top panel of the figure are boxes containing values for a variety of parameters. In the bottom panel are curves showing, in this instance, the influence of the time since last update on the number of AFVs killed by a Blue missile salvo. Separate curves appear for the first and second missiles in the salvo. As the user changes the parameter values by clicking the arrows next to the boxes, the curves will change to reflect the new parameter values. The user can also select any of the parameters to replace "Time of Update List" on the X-axis, or to replace "Missile" as the "key." The user can generate such a "results" display with any calculated quantity on the vertical axis.

Analytica's results display is good for a first look at a broad range of cases, but it does not allow one to bring to bear statistical methods and other data analysis techniques. For that we create an experimental design followed by a long sequence of runs covering the cases of interest and use Analytica's ability to export the results as a plain ASCII file. We can import the ASCII file into another software package (we chose Excel™) for further analysis. To illustrate, we ran cases for all 720 combinations of the parameter values shown in Table 6.1.

The parameter values in Table 6.1 reflect the Red vehicle configurations noted in the high-resolution simulations (see Chapter Five). But those runs considered much more dispersed Red formations than most military officers believe are likely.[3] Thus, we also

---

[3]Some officers have asserted that it is impractical to maintain command and control over a column of vehicles if the vehicles are separated by more than about 50 meters. They expect that if a column moves in a dispersed formation, vehicles will get lost along the way and will dribble in to the destination over an extended period of time. Once they begin arriving an enormous amount of time and effort will be required to re-form the unit.

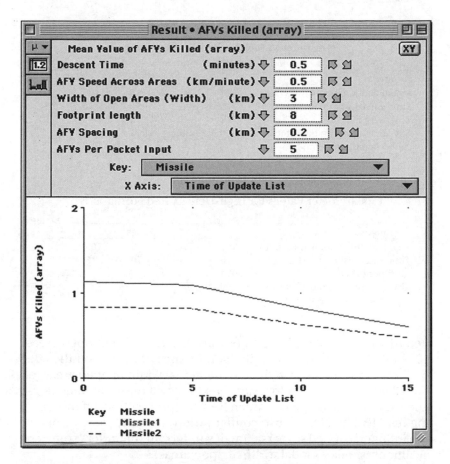

Figure 6.5—Analytica Results Display

considered the 20 combinations of parameter values in Table 6.2 as well, to reflect denser Red formations.

The baseline for our exploratory analysis is an attack against a very long column of closely and uniformly spaced (50-meter separation)

However, if dispersing his forces proves to be an effective enough countermeasure to long-range precision fires, Red might work out ways to do it. In part, then, one can consider the high-resolution cases as a test of whether Red has the incentive to explore this option.

Table 6.1

Parameter Values for Exploratory Analysis

Parameter	Values				
Modal width of open areas (km)	1	3	5	100	
Time since last update (min)	0	5	10	15	20
AFVs per packet	4	7	10		
AFV spacing within packet (m)	200	350	500		
Packet separation (km)	1	2.5	4	100	

Table 6.2

Parameter Values for High-Density Red Formations

Parameter	Values				
Modal width of open areas (km)	1	3	5	100	
Time since last update (min)	0	5	10	15	20
AFVs per packet	1000				
AFV spacing within packet (m)	50				
Packet separation (km)	0				

Red AFVs taking place in open terrain (modal width of open areas = 100 km). We will first examine the influence of terrain on the effectiveness of the attack by reducing the modal width of open areas to 5, then 3, and finally to 1 kilometer. Next, we will return to open terrain and examine the degree to which Red can protect himself by deploying his AFVs in a less dense configuration. Finally, we will consider the influence of Red dispersal tactics in terrain that offers some concealment (smaller modal width of open areas).

Attacks Against a Dense Red Column in Varying Terrain

Figure 6.6 shows the effectiveness of BAT against a very long column of closely spaced (50 meters) Red AFVs (see Table 6.2 for parameter values). Each salvo consists of two missiles. Each bar indicates a different type of terrain, which we represent by different modal lengths of open areas (W = 1, 3, 5, and 100 kilometers). The time since last update has no effect on these results because all parts of the Red column look the same. The rightmost bar represents our baseline, namely attacks against a long Red column of AFVs spaced uniformly at 50 meter intervals in terrain that offers no concealment. The

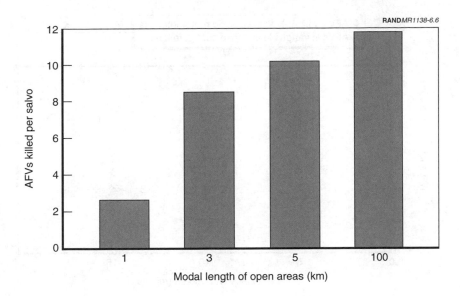

**Figure 6.6—Effectiveness of BAT Against an Infinite Column of AFVs
Uniformly Spaced at 50 Meter Intervals**

baseline effectiveness is just under 12 kills per salvo of two TACMS
missiles equipped with 13 BAT submunitions each.

We see from Figure 6.6 that modal length of open areas has some influence on kills per salvo, as Red vehicles are not vulnerable unless
they are in the open. However, the open areas must be quite small in
order to offer Red substantial protection.

Attacks in Open Terrain Against Varying Red Configurations

Red can protect his forces to some extent by dispersing them, even if
the terrain offers no concealment. Figure 6.7 shows kills per salvo as
a function of time since last update. All points are for a modal length
of open areas of 100 kilometers, which means the terrain
is entirely open and the Red targets are unobscured by foliage. The
top line consists of our baseline, in which the Red vehicles are in a

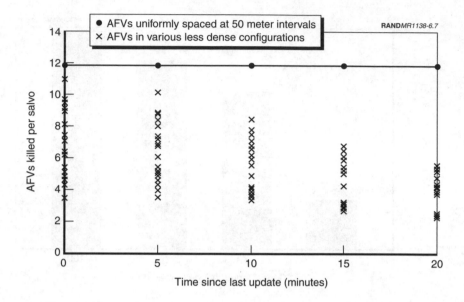

Figure 6.7—BAT Kills per Salvo in Open Terrain Against Various Red Configurations

continuous column, separated by 50 meters. All the remaining points assume various less dense configurations of the Red column, as generated from Table 6.1.

The time since last update has no effect on baseline kills per salvo because whenever the missiles arrive over the target, they will see some part of the Red column (it is very long, remember), and all parts look the same. But when Red configures his column differently, then the time since last update can matter. If Blue's missiles miss the packet they are aimed at, they may fall on the empty space between packets.

But there is another way of looking at Blue's effectiveness. As Red disperses his forces by increasing the average separation between vehicles, his column becomes longer. The time interval lengthens between the entry of the first vehicle into Blue's target area and the exit of the last vehicle. If Blue has limited stocks of ammunition, so

that the number of shots he can fire is independent of Red's exposure time, then Figure 6.7 gives the right view. If instead Blue is con-
strained by his maximum firing rate, Red may actually harm himself
by dispersing.

Figure 6.8 illustrates what happens if the depth of Blue's target area is
very small compared to the length of Red's column. For example,
1,000 AFVs separated by 50 meters and traveling at 75 kilometers per
hour will take only 40 minutes to pass a given point. But if the same
1,000 AFVs are configured in packets of 4 vehicles each, with vehicles
within a packet separated by 500 meters and packets separated by 4
kilometers (the lowest density configuration from Table 6.1), then
the column will require 20 hours to pass a given point. If Blue's firing
rate is limited (perhaps by C[4]ISR considerations), Blue can fire up to
30 times as many shots (if he has the ammunition to do so) and can

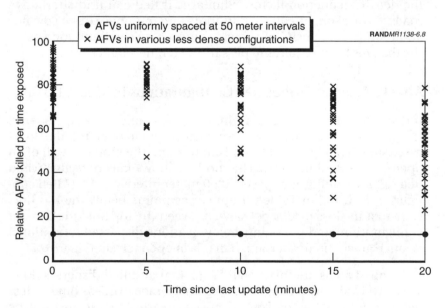

**Figure 6.8—Relative BAT Kills per Unit Time in Open Terrain Against
Various Red Configurations**

therefore kill more AFVs overall even though each salvo is much less effective.[4]

The solid line in both Figures 6.7 and 6.8 consists of the cases where the Red vehicles are in a continuous column, separated by 50 meters. The values are scaled so that the lines are at the same level in both figures. All the remaining points assume various different configurations of the Red column, as shown in Table 6.1.

We point out that in our treatment, the Red vehicles are restricted to traveling on roads. This makes the problem one dimensional, so that if Red reduces his density of AFVs, he increases the time his column takes to pass a point by the same factor. If Red vehicles could travel off road, they would have two dimensions in which to disperse, and the time used for maneuver would increase only as the square root of the density-reduction factor. Similarly, if Red can find alternative roads to travel on, he can split his force into multiple axes of advance and reduce the density of vehicles on each axis without increasing the time his force requires to pass through Blue's target area.

Attacks Against Various Red Configurations in Mixed Terrain

If Red is traveling through terrain that offers concealment, dispersal is a much more effective countermeasure than in open terrain. In all cases shown in Figure 6.9 (for mixed terrain), the modal length of an open area is 3 kilometers. The top line shows kills per salvo when Red AFVs are uniformly spaced at 50 meter intervals. All 27 Red configurations listed in Table 6.1 appear as points below the top line. Note that the loss in kills per salvo between the top line (the densest configuration) and the points below (various dispersed formations) is much greater in this terrain than it is in open terrain (Figure 6.7).

If the open areas are this small (W = 3 km) and Red disperses, Blue salvos will kill only one or two AFVs on average. Unless Blue is desperate (and has lots of TACMS missiles with BAT), it may not be

---

[4]This effect—i.e., this penalty to Red for dispersal—is particularly strong with respect to time vulnerable to Air Force sorties. Since sorties are typically prescheduled, without knowing when targets will be in the open, "firing rate" is limited. See also Ochmanek et al. (unpublished).

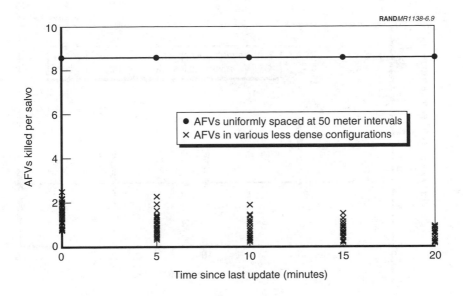

Figure 6.9—BAT Kills per Salvo in Mixed Terrain Against Various Red Configurations

worthwhile to fire. Decreasing the time since last update improves the effectiveness, though it still remains low.

As before, we rescale the data to estimate kills per time exposed instead of kills per salvo. The results appear as Figure 6.10. For a sufficiently long time since last update (or equivalently, a sufficiently large error in the impact time of Blue's missile), some dispersed configurations actually reduce Red's losses over the entire time the column is exposed, and not just per salvo. This reinforces the value to Red of dispersion coupled with concealment, and the value to Blue of accuracy.

In this instance, the penalty for dispersal is less and the relevance of the alternative view (kills per unit time) is also less. Long-range precision munitions such as TACMS with BAT are expensive, and Blue doctrine may call for firing only when he expects to achieve at least a specified threshold number of kills. According to Figure 6.10, Blue will have few opportunities to kill large numbers of AFVs per salvo in

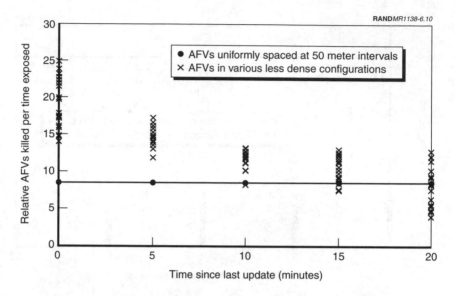

Figure 6.10—Relative BAT Kills per Unit Time in Mixed Terrain Against
Various Red Configurations

terrain with small open areas. Fewer firing opportunities translates to a slower rate of kills (kills per time exposed), even as it maintains weapon effectiveness (kills per salvo).

Afterword

The foregoing is intended as an illustration of the power of exploratory analysis and of the consequent utility of models small and agile enough to support it. We have constructed PEM, just such a small and agile model, and tested it against a high-resolution simulation model. Gaining confidence that PEM behaves more or less as the high-resolution model would, we have run hundreds of cases to help understand BAT's strengths and weaknesses.[5] Can it be de-

---

[5]The data for PEM 1.0 treats a relatively small-footprint precision weapon (sensor-fused weapons) as well as BAT. We choose to present results only for BAT because we have yet to compare results for the small footprint weapon with high-resolution results.

feated by terrain or by Red countermeasures? Does its effectiveness depend on accurate predictions of Red's future positions?

The analysis presented here is illustrative, not complete. But if we were to extend it, the understanding thus gained could be used to suggest improvements to the weapon or perhaps new weapons that have complementary strengths and weaknesses. PEM might also be used within a larger modeling structure to develop better doctrines for the use of the weapon under a variety of circumstances.

Chapter Seven

A SIMPLIFIED "REPRO MODEL"

APPROACH

Although PEM is a relatively simple model conceptually, it is represented by a stochastic computer program with analytical relationships distributed among many nodes. Thus, it lacks the simplicity of a closed-form analytical expression. Since some of the effects it represents are numerically more important than others, it is possible that a simpler expression would suffice for some purposes. It was therefore of interest to see whether we could reproduce reasonably well the predictions of PEM with something much simpler—essentially a relatively simple formula that could be written on the back of the proverbial envelope, in a trivial spreadsheet program, or in a small function within an operational-level model. This we call a Repro model. Engineers might think of it as a scaling relationship.

THE REPRO MODEL

PEM estimates the effects of terrain by simulating a Blue missile attack aimed to hit a Red AFV packet as it reaches a gap or open area in the otherwise closed terrain.

Red's Vulnerability Zone

The packet is not vulnerable in the entire gap. PEM describes a submunition descent time during which the Blue missile cannot acquire further targets. AFVs moving into the gap once the descent has started are not targeted and thus cannot be hit. Thus, Red's "vulnerability zone" is effectively reduced at the lower end by the distance Red's AFVs can move during the descent time. Further, if the footprint of the Blue missile is smaller than this zone, Red AFVs are only vulnerable in this footprint. Taking both cases into account, Red's vulnerability zone can be expressed as:

$$Z = \text{Min}[F, W - VT_d], \tag{7.1}$$

where W is the length of the open area; F, the diameter of the missile footprint; V, the speed of the Red AFVs in and around the open area; and T_d, the descent time of the submunition.

Blue's Targeting Error and the Vulnerability Function

Due to errors in estimating Red's speed—propagated over the time it takes to target and fire the missile and for the missile to fly to the target—Blue's missile may not reach the targeted gap at the same time as the targeted packet. PEM describes Blue's missile arrival time (compared to the arrival of the targeted packet, defined as t = 0) as a normal distribution whose standard deviation is the product of Blue's fractional error in estimating Red's in-terrain speed and the time since Blue's last update of the targeted packet's position and velocity.

Like PEM, the Repro model calculates the portion of each Red packet in the vulnerability zone as a function of the time the missile arrives. Depending on the length of the zone, the Red packet may overflow the zone, leaving some Red AFVs safe even when the packet is centered in the zone. Of course, if the packet fits entirely within the zone, the entire packet of N AFVs is vulnerable. The maximum number of AFVs from a single packet that can be vulnerable can be expressed as:

$$\text{VulnAFVs}_{\text{max}} = \text{Min}\left[N, \frac{Z}{S}\right], \tag{7.2}$$

where Z is the length of the vulnerability zone (cf., Equation 7.1), S is the spacing between AFVs within a Red packet, and N is the number of AFVs in each Red packet.

The number of vulnerable AFVs is at its maximum value if the packet is centered in the zone when the missile arrives. This value will hold so long as the zone contains the maximum number of AFVs (limited by the zone length or packet size, depending upon geometry). If we measure missile-impact-time t relative to when the targeted packet is centered in the open area, then the number of vulnerable AFVs decreases linearly when:

1. $t \geq [(Z - NS)/2]/V$ for NS < Z [the packet is smaller than the vulnerability zone]; or

2. $t \geq [(NS - Z)/2]/V$ for NS $\geq$ Z [the packet fills or exceeds the vulnerability zone].

The number of vulnerable AFVs continues to decrease until no AFVs from that packet remain in the zone. Figure 7.1 illustrates these processes for both Z > NS (left side) and NS >Z (right side).

In the (B) and (C) panels, the top row of circles shows packet location at time t = 0 and the lower row shows the location at the times when the packet moves away from maximum vulnerability, and when the packet becomes invulnerable, respectively. Since the process of entering a zone (and going from zero vulnerability to maximum

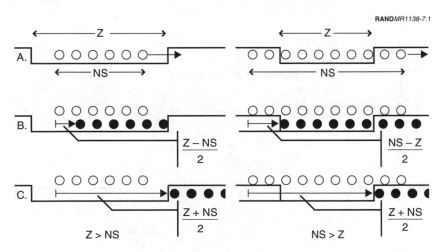

Figure 7.1—Geometry of Vulnerability Versus Time

vulnerability) is symmetric with the process of leaving a zone, the packet will be at maximum vulnerability for

$$T_c - \frac{|Z-NS|}{2V} < t < T_c + \frac{|Z-NS|}{2V} \qquad (7.3)$$

(where T_c is the time the packet is centered in the zone and V is the speed of the packet through the zone), and will have zero vulnerability otherwise. This vulnerability function is shown graphically in Figure 7.2.

Each Blue missile targets a particular packet, but because of the size of the footprint and the error in the missile impact time, we consider the possibility that it may kill vehicles in the leading and following packets, as well. Every packet's vulnerability function is identical in shape for a given zone, with each packet's function centered around the T_c for that packet. By definition, T_c for the targeted packet is zero, and T_c for the leading and following packet are the times those packets become centered in the zone. If the packets are separated (tail to head distance) by a distance P, then T_c values for the previous and subsequent packets relative to the targeted packet are

RAND*MR1138-7.2*

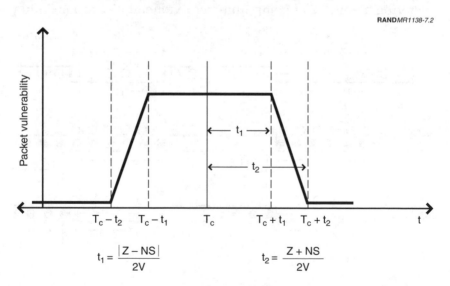

Figure 7.2—Vulnerability Function Versus Time

$$\frac{-(P+NS)}{V} \text{ and } \frac{(P+NS)}{V}, \text{ respectively.}$$

If we normalize the maximum value of the function to 1, the equations for the nonconstant regions of the function can be found by fitting to the boundary conditions at $T_c \pm t_1$ and $T_c \pm t_2$. The vulnerability function for a given packet, i, can thus be expressed as a function of time, like so:

$$\text{Vuln}_i(t) \approx \begin{cases} 0 & \text{for } t < T_{c,i} - t_2 \\[2mm] \dfrac{t + t_2 - T_{c,i}}{t_2 - t_1} & \text{for } T_{c,i} - t_2 \leq t \leq T_{c,i} - t_1 \\[2mm] 1 & \text{for } T_{c,i} - t_1 \leq t \leq T_{c,i} + t_2 \,. \\[2mm] \dfrac{-t + t_2 + T_{c,i}}{t_2 - t_1} & \text{for } T_{c,i} + t_1 \leq t \leq T_{c,i} + t_2 \\[2mm] 0 & \text{for } t > T_{c,i} + t_2 \end{cases} \qquad (7.4)$$

The missile arrival time is normally distributed about $t = 0$, so the probability of the missile's arriving at time t is

$$pr(t) = \frac{1}{\sigma\sqrt{2\pi}} e^{-\frac{1}{2}\left(\frac{t-\mu}{\sigma}\right)^2} = \frac{1}{\sigma\sqrt{2\pi}} e^{-\frac{1}{2}\left(\frac{t}{\sigma}\right)^2}, \qquad (7.5)$$

where σ is the standard deviation of the missile arrival time about μ, the mean. In this case, σ is the product of the fraction error in Blue's speed estimate and the time since Blue's last update of the Red packet's speed and position, and μ is zero.

By multiplying the vulnerability function for packet i by the probability function for the missile's arrival time and then integrating over all possible missile arrival times, we can calculate the expected value of that packet's vulnerability at missile arrival:

$$\langle \text{Vuln} \rangle_i = \int_{-\infty}^{\infty} \text{Vuln}_i(t) pr(t) dt. \qquad (7.6)$$

Recall that we earlier normalized the vulnerability function to achieve a maximum of 1. Multiplying this expected vulnerability by the maximum vulnerability of a packet yields the expected number of that packet's AFVs that are vulnerable when the missile arrives:

$$\langle \text{VulnAFVs} \rangle_i = \text{VulnAFVs}_{\max} \times \langle \text{Vuln} \rangle_i. \tag{7.7}$$

By doing a similar calculation for each packet, and then summing over packets, the expected total number of vulnerable AFVs seen by Blue's missile can be determined.

We will adopt a simplifying assumption that approximates the trapezoidal vulnerability function with a simple step function with the same undercurve area, as follows:

$$\text{Vuln}_i(t) \approx \begin{cases} 0 & \text{for } t < T_{c,i} - t_2 \\ y & \text{for } T_{c,i} - t_2 \leq t \leq T_{c,i} + t_2, \\ 0 & \text{for } t > T_{c,i} + t_2 \end{cases} \tag{7.8}$$

where

$$Y = \frac{t_1 + t_2}{2t_2}, \tag{7.9}$$

which is the ratio of the area of the trapezoidal vulnerability function (cf. Equation 7.4) to the area of the rectangular approximation described above.

For the ith packet, we can combine Equations 7.5, 7.8, and 7.9 to generate an expected vulnerability.

$$\langle \text{Vuln}_i \rangle = Y \left[\Phi \left(\frac{T_{c,i} + t_2}{\sigma} \right) - \Phi \left(\frac{T_{c,i} - t_2}{\sigma} \right) \right], \tag{7.10}$$

where $\Phi(x)$ is the normal cumulative distribution function (CDF):

$$\Phi(x) = \int_{-\infty}^{x} \left(\frac{1}{\sqrt{2\pi}} e^{-\frac{1}{2}u^2} \right) du. \qquad (7.11)$$

Though Φ cannot be expressed exactly in analytic form, a relatively simple approximation,[1] which is good to two decimal places, is:

$$\Phi(x) = \begin{cases} 0.0, & \text{for } x \le -2.6 \\ 0.01, & \text{for } -2.6 < x \le -2.2 \\ 0.1x(4.4+x)+0.5 & \text{for } -2.2 < x \le 0 \\ 0.1x(4.4-x)+0.5 & \text{for } 0 < x \le 2.2 \\ 0.99, & \text{for } 2.2 < x \le 2.6 \\ 1.0, & \text{for } 2.6 < x \end{cases} \qquad (7.12)$$

We added $\Phi(x)$ to the Analytica model as a library function for use in the Repro calculation.

Using Equations 7.2, 7.7, and 7.10, we can calculate the expected number of vulnerable AFVs from each packet. Summing over all three packets yields the expected total number of vulnerable AFVs encountered by Blue's missile. Because the function $\Phi(x)$ is symmetric around x=0, the vulnerability calculation for the leading packet yields exactly the same results as that for the following packet. Thus the result is

$$\begin{aligned} \langle \text{VulnAFVs} \rangle_{\text{total}} &= \text{VulnAFVs}_{\text{max}} \times \left(\langle \text{Vuln} \rangle_0 + 2 \langle \text{Vuln} \rangle_1 \right) \\ &= \frac{NZ}{Z+NS} \left[\begin{array}{l} \Phi\left(\dfrac{Z+NS}{2\alpha}\right) - \Phi\left(-\dfrac{Z+NS}{2\alpha}\right) \\ +2\Phi\left(\dfrac{2P+Z+3NS}{2\alpha}\right) - 2\Phi\left(\dfrac{2P-Z+NS}{2\alpha}\right) \end{array} \right] \end{aligned}, \qquad (7.13)$$

---

[1]Adapted from *Standard Mathematical Tables and Formulae*, CRC Press, Boca Raton, FL., 1991.

where Vσ has been rewritten as α. This is valid for all values of NS and Z. In the special case of $\alpha = 0$, there is no error in the arrival time of the Blue missile. In these cases, Blue's missile always achieves impact exactly when the targeted packet is centered in the open area. To ensure well-behaved results, Equation 7.14 is taken to be valid only for values of $\alpha > 0$. When $\alpha = 0$, the vulnerability of the targeted packet takes its maximum value (as described earlier, if the zone is longer than the packet, the entire packet of N vehicles is vulnerable; if the zone is shorter, only those vehicles that can fit inside the zone are vulnerable). The vulnerability of the leading and following packets can be readily calculated from the geometry of the zone. Between the tail of the leading packet and the head of the following packet exist two P intervals and the targeted packet of length NS. If the zone is less than 2P + NS in length, none of the nontargeted packets will be vulnerable. From this limit adding the lengths of the two packets in question makes it clear that for zones with lengths greater than 2P + 3NS, both nontargeted packets are fully vulnerable. In between these two zone lengths, some portion of the two packets will be at each edge of the zone. In this case as many AFVs will be vulnerable as can fit in the length Z – 2P. Thus, for $\alpha = 0$ the total vulnerability is:

$$\left\langle \text{VulnAFVs} \right\rangle_{\text{total}} \big|_{\alpha=0} = \begin{cases} \min\left[N, \dfrac{Z}{S} \right] & \text{for } Z \leq 2S_{\text{packet}} + NS \\[2ex] \dfrac{Z-2P}{S} & \text{for } 2S_{\text{packet}} + NS \leq Z \leq 2S_{\text{packet}} + 3NS \\[2ex] 3N & \text{for } Z > 2S_{\text{packet}} + 3NS \end{cases} \qquad (7.14)$$

Calculation of Expected Kills per Shot

Each missile has some maximum number of kills it can achieve, regardless of how many targets are present upon impact. The "actionable" number of vulnerable AFVs is the number of AFVs that the Blue missile can actually shoot at, given his kill limit (presumably due to targeting system or submunition limits). In terms of the Repro model's variables, this number is $N_{act} = M/FracKill$, where M is the maximum number of kills the missile can attain, and FracKill is the fraction of AFVs targeted by the missile that are killed.

As described earlier in Chapter Six, the number of AFVs seen by the incoming missile is approximately an exponentially declining function. The probability of the missile's finding a particular number of AFVs in the vulnerability zone upon arrival can be written as:

$$P(n) \approx Kr^n, \tag{7.15}$$

in which K and r are constants, and n is the number of AFVs seen. The expected number of vehicles seen (which is calculated in Equation 7.13 and 7.14[2]) can be expressed in these terms as:

$$\langle VulnAFVs \rangle_{total} \approx \sum_{n=0}^{\infty} n \times P(n). \tag{7.16}$$

Using Equation 7.16 and the fact that the probabilities sum to 1, and after some manipulation, we can calculate K and r as:

$$\begin{aligned} K &= \frac{1}{\langle VulnAFVs \rangle_{total} + 1} \\ r &= \frac{\langle VulnAFVs \rangle_{total}}{\langle VulnAFVs \rangle_{total} + 1} \end{aligned} \tag{7.17}$$

Using the probabilities from Equation 7.15, we calculate the expected number of actionable AFVs seen as:

---

[2]If $\alpha = 0$, the number of vehicles seen becomes deterministic (thus, the necessity of Equation 7.20). In this case the process described in this chapter is not necessary, and the expected kills value is simply $Min[M, AFVsVuln|_{a=0} \times FracKill]$.

$$\langle \text{AFVsAct} \rangle = \sum_{n=0}^{\infty} \text{Min}(n, N_{act}) \times P(n) \tag{7.18}$$

$$= \sum_{n=0}^{\infty} n \times P(n) - \sum_{n=N_{act}}^{\infty} (n - N_{act}) \times P(n)$$

$$= \langle \text{VulnAFVs} \rangle_{total} - Kr^{N_{act}} \sum_{n=N_{act}}^{\infty} (n - N_{act}) \times r^{(n-N_{act})}$$

$$= \langle \text{VulnAFVs} \rangle_{total} \times (1 - r^{N_{act}}).$$

Multiplying by the fraction killed we obtain the expected kills:

$$\langle \text{Kills} \rangle_{total,act} = \text{Frackill} \times \langle \text{VulnAFVs} \rangle_{total} \times (1 - r^{N_{act}}). \tag{7.19}$$

Calculation of the Terrain Effectiveness Multiplier

This process takes only terrain factors into account to calculate the number of kills a single Blue missile can be expected to achieve. Without these terrain factors, Blue's missile will always arrive on time (since there is no terrain to impede Blue's observation of Red's packets, and kills will not be limited to those AFVs in the open area but rather to all AFVs within the footprint of the missile. Thus, the vulnerable AFVs seen by the Blue missile in a "no-terrain" scenario are just those AFVs in the vulnerability zone at $t = 0$. Since there is no terrain, only the footprint of the missile limits the length of the vulnerability zone. Thus, the total number of vulnerable AFVs encountered by the Blue missile absent terrain is

$$\langle Vu\ln AFVs\rangle_{\text{no terrain}} =$$

$$
\begin{cases}
Min\left[N, \dfrac{F}{S}\right] & \text{for } F \leq 2P + NS \\[2ex]
\dfrac{F - 2P}{S} & \text{for } 2P + NS \leq F \leq 2P + 3NS \\[2ex]
3N & \text{for } F > 2P + 3NS
\end{cases}
\qquad (7.20)
$$

Taking into account the maximum kills each missile can achieve, the actionable number of vulnerable AFVs seen and kills achieved by the Blue missile in the no-terrain case can be written as:

$$
\begin{aligned}
\text{VulnAFVs}_{\text{no terrain, act}} &= Min[\text{Vuln AFVs}_{\text{no terrain}}, N_{\text{act}}] \\
\text{Kills}_{\text{no terrain}} &= \text{Vuln AFVs}_{\text{no terrain, act}} \times \text{Frackill}
\end{aligned}
\qquad (7.21)
$$

Finally, the multiplier to kills to be used in lower-resolution (than PEM) models is the ratio of Repro-calculated kills to the kills achieved absent terrain. The constant FracKill, present in both kill values, cancels, and the multiplier can be expressed as:

$$
\text{Multiplier}_{\text{terrain}} = \frac{\langle \text{Kills}\rangle_{\text{total, act}}}{\text{Kills}_{\text{no terrain, act}}}.
\qquad (7.22)
$$

COMPARISON OF RESULTS WITH THE PEM MODEL

To test how well this model agrees with the results of the PEM model, the Repro model was implemented in Analytica. Both models were input with identical datasets, the domains for which are shown in Table 7.1. Note that this compares expected values generated from the stochastic PEM model with deterministic values generated by RPEM. If stochastic variations were important in the context of a larger simulation, then a reprogrammed version of PEM rather than RPEM could be used as a subroutine.

Table 7.1

Dataset Used in the Validation of the Repro Model Results

Variable	Description	Units	Domain			
S	Spacing between AFVs	km	0.05	0.10	0.20	0.30
N	AFVs per packet	AFVs	5.0	15.0	100.0	
V	AFV velocity across open areas	km/min	0.25	1.25	1.50	
T_up	Time since last update	min	0	10	20	
W	Mean length of open areas	km	2	6	12	
T_d	Missile descent time	min	0.0	0.5	1.5	
P	Mean spacing between packets	km	1	3	6	
F	Missile footprint	km	2	8	12	
M	Maximum kills per missile	AFVs	6	12		
Frackill	Fraction of vuln. AFVs killed	—	0.25	0.7		

For each set of variable inputs, the outputs to be compared were:

- From PEM, the mean of a sample of 50 Monte Carlo iterations for the number of kills achieved by the first missile (from a possible salvo of two missiles); and

- From the Repro model, the calculated value of kills per shot. This value is deterministic, so no means need be taken.

The fit between PEM data and the Repro data was measured by fitting the data to the equation:

$$\langle PEMKills \rangle = m(ReproKills) + b, \tag{7.23}$$

using the least-squares method of regression. For the dataset described in Table 7.1, the results are recorded in Table 7.2.

These results suggest a reasonable fit for the Repro model with PEM. The calculated value of b indicates that both models predict zero kills

Table 7.2

Statistical Results from the Comparison
of the PEM and Repro Models

m = 0.94	Se(m) = 0.0017
b = 0.12	Se(b) = 0.0084
	R^2 = 0.90

in consistent ways, and the slope of 0.94 reveals that the Repro model's predictions are roughly on the same scale with those from PEM. Finally, the R^2 value of 0.90 means that 90 percent of the variation in expected-value kills on the PEM data can be explained by the Repro model, which is certainly reasonable for the level of resolution we are trying to achieve.

As mentioned earlier, the full PEM model has many inherent stochastic variations, which are not represented in the Repro model. Aside from the random arrival time of Blue's missiles relative to the centering of the targeted packet within the targeted open area, PEM's calculations for length of open areas and spacing between Red AFV packets are varied randomly around input values. This leads to stochastic variation in the results generated by PEM that is not reflected in the Repro model's deterministic results (although the distribution of the arrival time for Blue's missile is taken into account in the derivation of the Repro model).

Figure 7.3 illustrates the random nature of PEM by plotting, for a representative "setting" of various input parameters, the standard deviation of PEM's calculation of the number of AFVs killed by a given salvo of "big missiles."[3]

As expected, the variance of PEM's results increases as the Time of Last Update (and therefore the variance in the arrival time of Blue's missile) increases. Notice, though, that some variance exists even when no error exists in Blue's missile arrival time. This reflects the uncertainty associated with the spacing between Red's AFV packets and size of the targeted open area.

Figures 7.4 and 7.5 give an even clearer picture of the variation in AFVs killed per salvo. Figure 7.4 shows the distribution when there is no error in Blue's missile arrival time, and Figure 7.5 shows the same when Blue's missile arrival time has a distribution that results from a time of last update of 15 minutes.

---

[3]The values used in this example are: Spacing between AFVs, 0.2 km; AFVs per packet, 15; AFV velocity across open areas, 1.25 km/min; Mean length of open areas, 6 km; Missile descent time, 0.5 min.; Mean spacing between AFV packets, 3 km; Missile footprint, 8 km; Maximum number of kills per missile, 6; and Fraction of vulnerable AFVs killed, 0.25. Two hundred and fifty Monte Carlo draws were made.

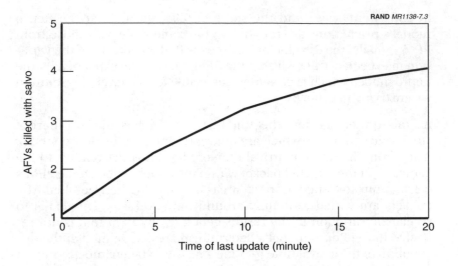

Figure 7.3—The Standard Deviation of AFVs Killed with
Salvo for Various Values of Time of Last Update

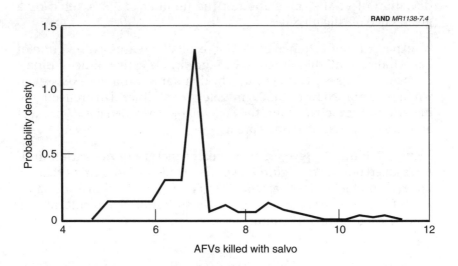

Figure 7.4—Probability Density (in %) of AFVs Killed with
Salvo, Given No Error in Arrival Time

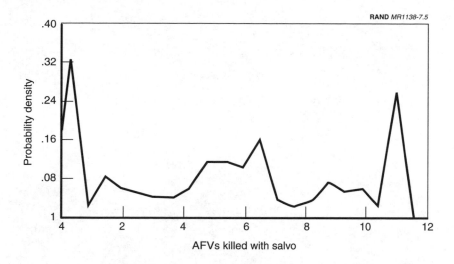

Figure 7.5. Probability Density (in %) of AFVs Killed with Salvo, Given a Time of Last Update of 15 Minutes

It is easy to see that there is a degree of stochastic richness in PEM that is not captured by the RPEM calculation. Nevertheless, RPEM appears to be rather accurate in reproducing PEM's mean (or "expected value") results. Thus, we conclude that so long as one needs only expected kills, and not a probability distribution of kills, it is reasonable to use the Repro model in lower-resolution models such as EXHALT to represent the effects of terrain on the effectiveness of Blue's air and missile efforts against a Red AFV advance.

Chapter Eight

CONCLUSIONS

RECALLING OBJECTIVES

At the start of this report, we gave our purpose as describing analytically how the effectiveness of long-range precision weapons should be expected to vary when used against a moving armored column, depending on variables usually treated only in much more complex models. We considered:

- Characteristics of the Blue C^{4}ISR system;
- Characteristics of the Blue weapon systems;
- Maneuver pattern of the advancing Red armored column;
- Terrain features; and
- Blue tactics, including salvo offsets.

OBSERVATIONS

These factors can make a huge difference in the effectiveness of long-range precision fires, as measured by kills per salvo or sortie. Consider the "standard" target for such fires to be a column of armored vehicles separated randomly but typically by no more than 50 meters, traveling across terrain that offers no concealment (e.g., the desert). Long-range precision fires can be extremely effective against such a standard target. When the above factors are varied, however, kills per salvo may be reduced by one or two orders of magnitude.

Moreover, these factors interact. One cannot model the influences of these factors in a simple way—for example, as a product of a terrain adjustment, a Red dispersion adjustment, a C⁴ISR adjustment, and so on. The sensitivities of outcome to one factor depend strongly on other factors, and linear sensitivity analysis around some baseline would be misleading. Thus, we need to use exploratory analysis to characterize how the factors combine to dictate Blue success or failure.

The Red maneuver pattern, which can reflect a passive Red countermeasure against the Blue attack, interacts with several of the other factors. Changing the Red maneuver pattern has a much greater effect in terrain with small open areas than in terrain with large ones. Also, changing the maneuver pattern (particularly the AFV spacing within a packet) is more important for weapons with small footprints (SFWs) than for those with large footprints (ATACMS/BAT). However, a weapon with a large footprint loses its advantage over a weapon with a small footprint in terrain with no large open areas to shoot at.

Depending on the other factors, the time since last update (a primary C⁴ISR parameter) can range from very important to completely irrelevant. In PEM, the time since last update influences the results through its effect on the time-of-arrival error of the Blue missile, and NOT through an effect on a spacial error in the impact point. If Red maneuvers in long columns of uniformly spaced vehicles, then kills per salvo are the same regardless of the time since last update because the column looks the same regardless of when the missile arrives. But if Red changes his maneuver pattern, it becomes important for Blue to place his shots on target precisely when the Red packet arrives. Accurate timing of shots becomes even more important if open areas are small, at least in the sense that kills per shot decline by a larger fraction as time since last update increases. But if open areas are small enough, kills per shot may be too meager for shooting to be worthwhile even if the time since last update is zero, especially if weapon-descent time is long.

If the Red formation maintains strict discipline and moves with constant speed, then prediction will be easier. A small time since last update will pay dividends. But if Red is less organized or deliberately changes speeds, effectiveness may be low even with short times from last update, especially in mixed terrain.

In addition, the ability to discriminate between live and dead targets is important if multiple shots were fired into the same area without leaving time for dead targets to stop and cool or if kills per weapon is high.

Offsetting missiles in time often has only a marginal effect by reducing the likelihood of a second missile attacking a portion of a packet or packet group that has already been depleted. Offsetting the impact point has less effect for large-footprint weapons. In contrast, offsetting SFWs is quite important if they are employed, as suggested by Ochmanek et al. (1998), to "annihilate" the leading edge or nose of an advancing column. Otherwise, dead-target effects (failure to discriminate) would greatly reduce effectiveness. This, however, should be relatively straightforward operationally.

Some subtleties of outcome cannot reasonably be predicted from a PEM-level model. Here are some examples we have noted from high-resolution simulation:

- When ATACMS/BAT is employed against columns at crossroads, kills per shot can be sensitively dependent on the orientation of the columns and roads, and on the weapon logic used. In some instances, the weapon logic is confused by the pattern of signals and the submunitions are laid down on lines intermediate between good lines on the ground containing targets.[1]

- Weapon effectiveness can be sensitive to the level of acoustic noise due to vehicles in the general area, as well as the impact on directional signal-to-noise ratio of terrain. Thus, one cannot calibrate a PEM-like model by simply using a high-resolution model against a single target and then extrapolating. PEM should be considered better for scaling than for exact prediction.

---

[1]Initial data analysis suggested that this effect might be a major factor in the overall low effectiveness of weapons in the high-resolution simulations. Subsequently, we have come to believe that microscopic urban clutter, as discussed in earlier chapters, also played an important role in attacks against crossroads.

- Many details of operations affect results to some extent. For example, columns may show rigid discipline or may show highly random spacings among vehicles and units.

We have not addressed the issue of Blue doctrine for long-range precision fires, except for the matter of offsetting weapons. Under what circumstances should Blue even attempt to stop an armored column by long-range fires? What weapons should be used against large versus small groups of Red vehicles, or against armored vehicles versus trucks (if Blue's surveillance is capable enough to distinguish vehicle types)? As a stand-alone model, PEM is not suited for examining Blue doctrine, but as a quick and efficient subroutine embedded in a larger model, it may be. The larger model would generate groups of Red vehicles crossing the chosen terrain, and PEM could quickly evaluate the potential of each of several Blue weapons against all the groups as they crossed the various open areas. A wide variety of doctrines could be constructed and evaluated in such a modeling environment.

One might immediately assume that the larger model would be a high-resolution simulation model that tracks the movement of thousands of individual Red vehicles along specific roads in a terrain represented by millions of cells in an XY plane. But this need not be the case. The larger model could instead sample potential target groups from statistical distributions. Such a statistical model might be developed much as PEM was and then calibrated to a high-resolution simulation.

COMPARING WEAPON TYPES

Most of our study has focused on weapons comparable to ATACMS/BAT, but we also considered air-delivered sensor-fused weapons using some performance figures cited in an earlier study (Ochmanek et al., 1998) for the kills per sortie to be expected from an F-16 with four sensor-fused weapons attacking targets in the open. Several observations from applying PEM are of interest here. First, time from last update is much more important for SFWs than for ATACMS/BAT when attacking targets in the desert. Figure 8.1

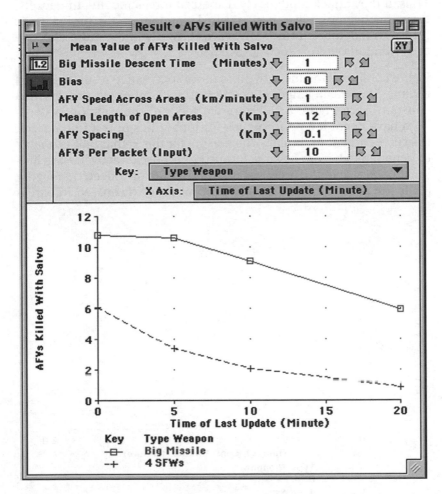

NOTE: Assumes a 25% error in estimating invaders' average speed along the road
between the time of observation and the intended time of weapon impact.

Figure 8.1—Sensor-Fused Weapons Versus BAT in Open Terrain

illustrates this and is qualitatively consistent with results from high-
resolution simulation accomplished for a 1996 DSB study
(Matsumura, Steeb, Herbert, et al., 1997). The reason behind this,
simply, is that the sensor-fused weapons have small footprints.

Thus, if they attack moderately dispersed forces moving in packets, their effectiveness will be sensitively dependent on the timing error. It follows that direct-attack SFWs are currently much more effective than standoff versions such as JSOW.[2,3]

Figure 8.2 shows an illustrative and more speculative result for mixed terrain (3 km mean length of open areas, with other parameters held constant from Figure 8.1). Here we see that the weapon systems both have degraded performance for long times of last update, but the curves are more similar. The reason, for this is that BAT is unable to benefit much from its large footprint because effectiveness is limited by the size of the open areas. Another factor reflected in Figure 8.2 is the effect of BAT's larger "descent time" discussed in earlier

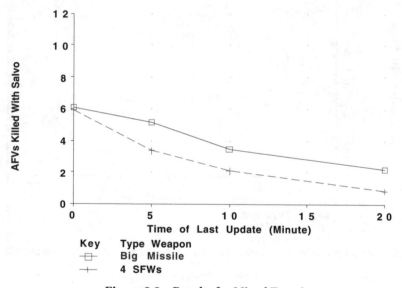

Figure 8.2—Results for Mixed Terrain

---

[2]See also a forthcoming RAND study for the Air Force by Edward Harshberger and colleagues. That study examines such weapon issues in considerable detail using high-resolution simulation.

[3]The display shown in Figures 8.1 and 8.2 refers to "Big Missile," by which we mean missiles akin to ATACMS with BAT.

chapters. Such considerations suggest that cost-effective comparisons, when assessing alternative mixes of weapons, need to be conducted quite carefully.

SUMMARY EFFECTS: LOOKUP TABLES FOR USE IN OTHER MODELS

So far, we have described our understanding of phenomenology,
a relatively detailed desktop model, and a simplified "repro" model.
Tables 8.1 and 8.2 go farther and summarize results of many
thousands of PEM runs in estimating the effectiveness of an
ATACMS/BAT-like system and an F-16/SFW system versus the
attacker's choices of AFV spacing, the type of terrain, and the time
from last update for the interdictor's weapon system. Since actual

Table 8.1

Kills per ATACMS/BAT Salvo for 0, 10, and 16 Minutes from Last Update[a]

Dispersal/Terrain	Open	Mixed	Primitive Mixed
No Timing Error			
Very tight	12	10	1
Dispersed	11	6.0	0.2
Very dispersed	6.2	2.9	0.06
10 Minute Errors			
Very tight	12	10	1
Dispersed	9.1	2.2	0.3
Very dispersed	4.9	1.8	0.15
16 Minute Errors			
Very tight	12	10	1
Dispersed	6.0	3.4	0.23
Very dispersed	3.1	1.1	0.15

Definitions: Very tight: 50 meters per AFV, 100 AFVs
per packet; Dispersed: 100 meters per AFV, 10 AFVs per
packet; Very dispersed: 200 meters per AFV, 5 AFVs per
packet. Open: 12 km open-area mean widths; Mixed: 3
km open-area mean widths; Primitive: 1 km open-area
mean widths.

[a]These figures assume a factor of 0.25 for the fractional
error in estimating maneuver speed along the road.
Actual time errors in projecting the arrival of targets in
an open area are then four times the numbers shown
in the tables.

effectiveness numbers would also depend on details not reported here (e.g., the acoustic environment due to the particular types of vehicles in the march, their configuration, and their interaction with the environment), what matters most is the relative numbers. As can be seen by comparing the top-left and bottom-right figures, we should expect a factor of nearly 100 in weapon effectiveness as a function of these three variables.[4]

Table 8.2 shows analogous results for SFWs, if used in a manner similar to that described in Ochmanek et al. (1998).

METHODOLOGICAL CONCLUSIONS

Without elaboration, let us merely observe that the study reported here demonstrates concretely the feasibility and power of an approach to analysis that actively works the gamut from high-resolution, entity-level, man-in-the-loop simulation on the one extreme, to exploratory analysis with fast-running desktop models on the other.[5] One of our principal objectives in undertaking this work was to accomplish such a demonstration. When such analytic work is used to inform and exploit empirical work, including large-scale field experiments, a great deal can be learned about the phenomenology of future military operations—including the risks associated with them and how to mitigate those risks.

---

[4]By far the most effective way to use PEM in understanding the situational and tactical effects is by working with PEM interactively. A single "exploratory analysis" session, possible even by a nonprogrammer, can be quite illuminating. Much is lost when we abstract results for the print media.

[5]For our prior discussions on the matter advocating such an approach, see Davis, Gompert, Hillestad, and Johnson (1998) and Davis, Bigelow, and McEver (1999). For theory underlying the modeling issues, see Davis and Bigelow (1998).

Table 8.2

Kills per F-16 Sortie with Sensor-Fused Weapons for 0, 10, and 16 Minutes from Last Update[a]

Dispersal/Terrain	Open	Mixed	Primitive Mixed
No Timing Error			
Very tight	8.5	8.4	4
Dispersed	6.1	6.0	2.7
Very dispersed	2.7	2.6	1.0
10 Minute Errors			
Very tight	8.5	8.4	4.0
Dispersed	2	2.1	1.2
Very dispersed	0.95	0.95	0.65
16 Minute Errors			
Very tight	8.5	8.4	4
Dispersed	0.9	0.86	0.54
Very dispersed	0.36	0.36	0.25

Definitions: Very tight: 50 meters per AFV, 100 AFVs per packet; Dispersed: 100 meters per AFV, 10 AFVs per packet; Very dispersed: 200 meters per AFV, 5 AFVs per packet. Open: 12 km open-area mean widths; Mixed: 3 km open-area mean widths; Primitive: 1 km open-area mean widths.

[a]These figures assume a factor of 0.25 for the fractional error in estimating maneuver speed along the road. Actual time errors in projecting the arrival of targets in an open area are then four times the numbers shown in the tables.

RAND'S HIGH-RESOLUTION FORCE-ON-FORCE MODELING CAPABILITY[1]

OVERVIEW

RAND's suite of high-resolution models, depicted in Figure A.1, provides a unique capability for high-fidelity analysis of force-on-force encounters. In this suite the RAND version of Janus serves as the primary force-on-force combat effectiveness simulation and provides the overall battlefield context, modeling as many as 1,500 individual systems on a side. The Seamless Model Interface (SEMINT) integrates Janus with a host of other programs into one coordinated system, even though the participating models may be written in different programming languages running on different hardware under different operating systems. In effect, SEMINT gives us the ability to augment a Janus simulation by specialized high-fidelity computations of the other partaking models, without actually modifying the Janus algorithms.

As currently configured, Janus conducts the ground battle, calling on the RAND Target Acquisition Model (RTAM) to provide more accurate calculation of detection probabilities of special low-observable vehicles. The Model to Assess Damage to Armor by Munitions (MADAM) simulates the effects of smart munitions, including such aspects as chaining logic, multiple hits, and unreliable submunitions, while the Acoustic Sensor Program (ASP) provides a detailed

---

[1]This appendix is abstracted from informal documentation provided to us by colleague Tom Herbert.

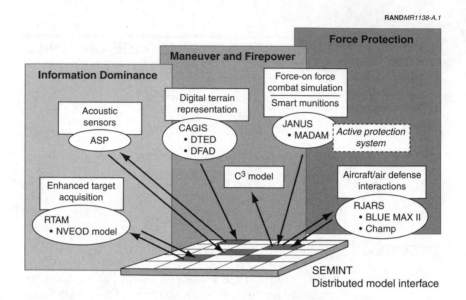

RAND*MR1138-A.1*

Figure A.1—RAND's Suite of High-Resolution Models

simulation of acoustic phenomenology for such systems as air-delivered acoustic sensors and wide-area munitions. Should the conflict involve helicopter or fixed-wing operations, the flight planners BLUE MAX II (fixed wing) and CHAMP (helicopter) determine flight paths for the missions, flown against the actual Janus threat, and RAND's Jamming and Radar Simulation (RJARS) conducts the defense against the aircraft, including detection, tracking, jamming, and SAM operations. The Cartographic Analysis and Geographic Information System (CAGIS) provides consistent geographic information to all the simulations, while SEMINT passes messages among the models and maintains a Global Virtual Time to keep the models in synchronization.

SCENARIOS

RAND makes use of Standard High-Resolution scenarios, made available by U.S.-TRADOC Analysis Center (TRAC), and modifies them as necessary to meet individual project objective needs. When suitable standard scenarios are not available, or necessary modifications to existing scenarios are too extensive to be practical, scenarios or vignettes are developed at RAND to isolate and examine essential elements of analysis (EEA) identified for individual projects. An appropriate level of awareness to the validity of each scenario with respect to likely "real-world" situations and contingencies is maintained, and assumptions are always based on "best available data." Vignettes are thoroughly gamed and then meticulously scripted to ensure "reasonable" tactics and behavior in the absence of human reaction and intervention when running in the batch mode.

Although Janus affords the capability of modeling division-versus-division level engagements, typical vignettes are developed at the battalion task force-versus-brigade, or brigade-versus-division level. Vignettes are normally scripted to simulate 60 minutes or less of real time. In batch mode, the model suite typically runs at or faster than real time, depending upon the complexity of the vignette. (It can also be run interactively, with Red and Blue gamers.) Each vignette is iterated (nominally) 30 times to obtain a reasonable sample, and the resulting statistics are analyzed both aggregately and by iteration.

POSTPROCESSOR

To analyze the output of the high-resolution suite, RAND has developed a postprocessor. It is written in SAS™ (the Statistical Analysis System) to take advantage of the enormous sorting, ordering, manipulative, and computational power offered by that software when dealing with prohibitively large, free-form datasets. The software also offers a push-button type interface for standard options programmed in SAS. This offered as close to an ideal solution as could reasonably be expected for the large datasets for each excursion in very large analytic matrices associated with Janus and its associate models.

The postprocessor displays data in a variety of forms, from simple tables to line graphs to pie charts, to bar and stacked bar charts, to complex, three-dimensional plots necessary for spotting trends in extremely large output datasets. It also prepares data for plotting on terrain maps in order to spot spacio-temporal relationships. These graphic displays use varying icons and colors to represent large numbers of different parameters in a single display. For example, one color may represent a battlefield system that was detected but not engaged, while another may represent a system that was engaged but not killed, while another may represent a system that was killed by indirect fire, while yet others represent systems that were killed by various direct-fire weapon systems.

The postprocessor has continued to evolve as new insights from a wide-ranging variety of studies have generated new and innovative ways of viewing and presenting data from high-resolution simulations. Each time a new technique for viewing the data is developed, it becomes an integral part of the postprocessor as a new push-button option.

PEM AND THE HIGH-RESOLUTION MODELS

Only a subset of the high-resolution models is directly involved in simulating the phenomena represented in PEM, namely the effect of long-range precision fires against a specified group of target vehicles. Janus simulates the movement of the Red vehicles. From the Janus output, therefore, PEM obtains the Red march doctrine parameters, including the number of vehicles per packet, the separation of vehicles in a packet, the separation of packets, and the velocity of the Red vehicles (see Appendix B). CAGIS models the terrain, providing PEM with information on the lengths of open areas (see Chapter Five). MADAM calculates the effects of long-range fires against groups of Red vehicles (see Appendix C). SEMINT coordinates the other models.

Other high-resolution models are indirectly involved in the simulation of long-range precision fires. The DSB '98 cases from which we took our data involved a man-in-the-loop who decided the aim points and impact times of the long-range fires. He based his decisions on the simulated results of surveillance from long range by unmanned aircraft, and in different cases he received information of varying completeness. But PEM does not address the problem of deciding when or at what to shoot, so important as this aspect of the simulation is in determining the overall effectiveness of long-range precision fires, it is not directly relevant to PEM.

MADAM

For PEM, the key high-resolution model is the Model to Assess Damage to Armor by Munitions (MADAM). Figure A.2 illustrates its operation.

MADAM was originally written by the Institute for Defense Analysis (IDA). RAND has added significant additional capability in the form of upgrades capable of modeling the technologies associated with the following munitions:

- Seek And Destroy ARMor (SADARM)
- Sensor-Fused Weapons (SFW-Skeet)
- Damocles
- Low-Cost Anti-Armor Submunition (LOCAAS)
- Terminally-Guided Weapon/Projectile (TGW/TGP)

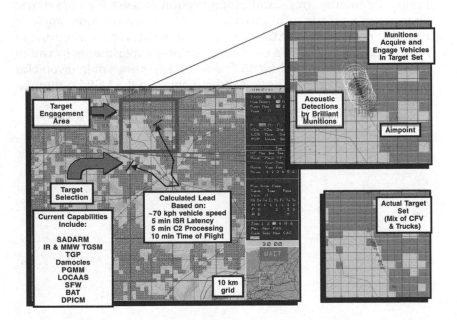

Figure A.2—Operation of MADAM

- Precision Guided Mortar Munition (PGMM) (Infra-Red (IR) & Millimeter Wave (MMW))
- Brilliant Anti-Tank (BAT)
- Wide Area Munitions (WAM).

The model provides a capability for simulating and analyzing chain logic, false alarm rates, hulks, submunition reacquisition, shots, hits and kills, as well as bus, munition, and submunition reliability. For example, to estimate how many vehicles are killed by a BAT, MADAM simulates the separation of the bus from the launch vehicle, the separation of submunitions from the bus, several stages of acoustic seeking and deployment by the submunitions as they descend, an IR detection stage, and a final shot/hit/kill event for each submunition. The outcome at each stage is determined, in part, by a random draw.

MADAM exists as both a stand-alone model and a subroutine of Janus. Ordinarily, the stand-alone version is used for parametric analyses as a precursor to provide focus for force-on-force analytic runs that draw on the MADAM version that resides as a subroutine in Janus. For this paper we used it to perform experiments in which salvos of one or two TACMS/BAT were fired at groups of Red vehicles of sizes and configurations that did not occur in the DSB '98 simulations.

RED'S MARCH DOCTRINE AND THE VARIABLES N, S, P, AND V

In this appendix we discuss how we calibrated four parameters in PEM from the high-resolution simulations. The parameters are: N, the number of AFVs in a packet; S, the separation of AFVs within a packet; P, the separation between successive packets; and V, the speed at which packets cross open areas.

In the DSB '98 simulations, there were 13 different types of Red vehicles, as shown in Table B.1. Of the 543 Red vehicles in the simulation, 104 were AFVs, consisting of the four types RADAXX, RAPC1F,

Table B.1

Numbers of Red Vehicles by Type

Vehicle Type	Number
RADAT	12
RADAXX**	4
RAPC1F**	35
RAPCXF**	32
RHELR	36
RLCVA	12
RLCVAA	27
RLCVM	12
RLTATK	7
RMCMV	21
RMCVGM	12
RTANKF**	33
RTRK3	300
Total	543

** AFVs.

RAPCXF, and RTANKF. But the other types also sometimes appeared in the footprint of a TACMS salvo.

The Red force split into three columns, following the routes shown in Figure 5.1. A detailed examination of the southern column at two simulated times (t = 25 minutes and t = 100 minutes) reveals that it consists of three kinds of packets, plus some miscellaneous vehicle that don't appear to be grouped into packets. Figure B.1 shows the distances the packets have traveled along the road from a common reference point. The three columns at the left show packet positions 25 minutes into the simulation, while the three columns to the right show the positions of the same packets at 100 minutes. Packets of different kinds are shifted horizontally in the figure so the reader can distinguish them. In the actual simulation, all packets traveled along the same road. It was presumed that the roads were wide enough to accommodate parallel columns.

The first kind of packet might be called a reconnaissance packet. Packets consisted of three vehicles, usually one each of RMCMV, RMCVGM, and RLCVAA. Vehicles within a packet were separated by

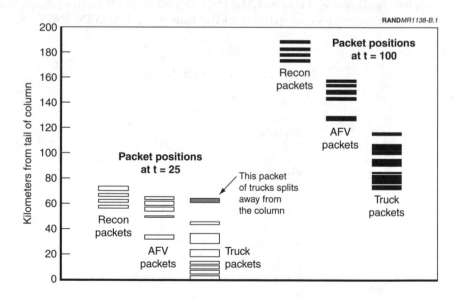

Figure B.1—Relative Positions of Packets in the Southern Column

an average of about 700 meters, and the spacing between packets was 3.2 to 4 km. We identified the same four packets at both 25 and 100 minutes into the simulation; they had the same vehicles, vehicle separations, and packet separations at both times. These packets moved forward in lockstep, therefore, at a speed of 94 km/hr.

The second kind of packet consisted of 3 to 10 (average 6.7) AFVs. Because AFVs were our primary interest, we examined AFV packets in the other two axes of advance (center and north) as well, and taking them all together, they contained 3 RADAXX, 21 RAPC1F, 31 RAPCXF, and 32 RTANKF (a total of only 87 of the 104 AFVs in the simulation). The average separation of AFVs in a packet was 350 meters, thought the separation varied widely (150–600 meters). The gaps between packets were typically 1 to 3.8 km. Sometimes, however, there were very long gaps. For example, the last packet of AFVs in the southern column lagged 14 km behind the others. Comparing the vehicle positions at times 25 and 100 minutes shows these packets were moving forward in lockstep at 76 km/hr.

A third kind of packet consisted of trucks (vehicle type RTRK3). The packets we have looked at (southern column only) contained from 7 to 18 vehicles, with 10 vehicles being most common. The separation of trucks within a packet averaged about 300 meters, and the gaps between successive packets averaged 3.9 km. The forwardmost truck packet in the column at t−25 minutes split from the column before t=100 and took a more southerly route. But the remaining seven packets moved forward in lockstep at 58 km/hr.

There were 123 vehicles in the three kinds of packets in the southern column. Another 22 vehicles traveled alone or in pairs, and are not shown in Figure B.1.

Packets of different kinds traveled at different speeds, so the order of the packets changed over time. At t=25 minutes, the chain of reconnaissance packets overlapped the chain of AFV packets, which in turn overlapped the chain of truck packets. But at t=100 minutes, the reconnaissance packets had all pulled ahead of the lead AFV packet, and the last AFV packet had passed the lead truck packet. Thus, the column, which began as a mixture of all the vehicle types, over time sorted itself into three more nearly homogeneous sections.

Historical rates of advance of mechanized forces facing even light opposition have only occasionally exceeded 40 km/day (Helmbold, 1990). Thus, Red can follow the march doctrine described here only a small fraction of the time, an hour or less per day on the average. The remainder of the time his force must be stationary (faster vehicles would be stationary somewhat longer so the slower vehicles could catch up). Of course, this report addresses only the problem of attacking the Red force while it is moving. The problem of detecting and attacking Red vehicles at rest is outside our present scope.

AFVs KILLED PER SALVO

The high-resolution model (MADAM, used within the larger suite of models described in Appendix A) simulates in detail the process by which BAT submunitions search acoustically for target vehicles and then home in on them by infrared sensing. Each submunition and each Red vehicle is simulated individually. In PEM, by contrast, we assume a missile salvo kills a fixed fraction (FracKill) of the Red vehicles in the footprint, so long as the number of Red vehicles is relatively small and the limited number of submunitions does not become a factor. For a salvo of two TACMS missiles, we set FracKill = 0.41, which we scaled down to about 0.22 for a single missile to account for redundant targeting effects. Figure C.1 shows results from the DSB '98 simulations, overlaid with the aggregate relation we have used in PEM. The straight line used ignores some points on the right side of the figure because we concluded that those reflected unusual cases where AFVs in two or more aim areas were being lumped together artificially by the data-analysis procedure.

When too many AFVs are in the missile's footprint, the number of kills may be limited by the number of BAT submunitions (each TACMS/BAT carries 13 submunitions). To account for this, we truncated kills at a maximum value, MaxKills. Too few TACMS salvos in the DSB '98 simulations targeted large numbers of AFVs to allow us to confidently estimate the maximum kills per salvo, so we based the estimate on additional experiments with the high-resolution simulation. In these experiments we fired salvos of one and two TACMS at groups of target vehicles ranging from 1 to 36 vehicles per group. The vehicles were arranged in different configurations, as if they were traveling in single file or meeting at a crossroad or at a fork in a

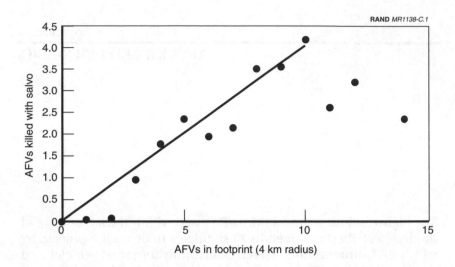

Figure C.1—Kills Versus Targets

road. A single TACMS/BAT missile killed roughly 5 target vehicles out of a group of 18, 6 out of 27, and 7 out of 36. Two missiles killed about twice the number of vehicles as one. We therefore truncated PEM's kill per salvo at MaxKills = 6 for a salvo of one missile. This implies a limit of 12 kills per salvo of two missiles for the conditions assumed (especially the acoustic environment).

The points in Figure C.1 are actually averages of many trials from the high-resolution model. The point with 5 AFVs in the footprint, for example, is the average of 155 trials. There is a considerable amount of variation in the number of AFVs killed per salvo, as shown in Figure C.2. To represent this variation, we make the parameter FracKill in PEM a random variable with a triangular distribution. Its minimum, mode, and maximum are (for a two-missile salvo) 0, 0.41, and 0.82, respectively.

DISCUSSION

For salvos with three to ten AFVs in the footprint, there is a reasonably good match between kills per salvo calculated from PEM's and

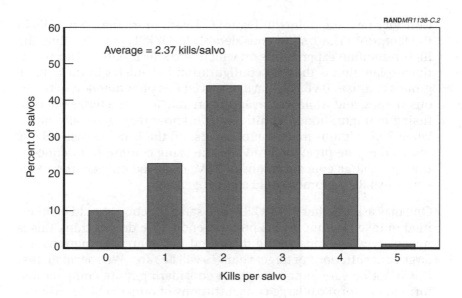

Figure C.2—Variation in Kills per Salvo
(Includes Only Salvos with 5 AFVs in Footprint)

kills per salvo from the high-resolution simulation. But when the number of AFVs in the footprint is either very small (one or two) or larger than ten, PEM appears to estimate significantly more kills per salvo than the high-resolution model.

Consider first the points in Figure C.1 with only one or two AFVs in the footprint. In the DSB '98 simulations, almost no AFVs were killed in these trials. However, we also conducted some experiments with the high-resolution model in which a single AFV was killed by a single TACMS with probability 0.87, even when the missile's aim point was as much as six kilometers distant from the AFV. On the face of it, the high-resolution model is contradicting itself! However, a second experiment showed that when background noise was present, BAT could no longer pick up a target group of vehicles from as far away. Perhaps when the target group contains only one or two AFVs, it produces too little noise to stand out from the background except when BAT's aim point is very close to the target group. Whether this explanation resolves the apparent contradiction awaits further investigation.

Consider next the points in Figure C.1 with more than ten AFVs in the footprint. These show considerably fewer kills per salvo than the high-resolution experiments on which MaxKills is based. Our tentative explanation is that the configuration of vehicles in these large groups confuses BAT. We have observed this phenomenon in previous studies, and while we haven't been able to characterize the confusing configurations definitively, we know they generally occur when Red columns meet at crossroads. In the DSB '98 cases, Red's march doctrine precluded AFVs in the same column from concentrating, so large concentrations of AFVs occurred only at crossroads where two Red columns could come together.

One may ask whether PEM's kills-per-salvo function should be modified to take account of this phenomenon. We do not think this is necessary when considering dispersed Red march formations, for large concentrations of target vehicles will be rare. We speculate that it will not be very important when considering more compact Red formations, for most large concentrations of target vehicles will not occur at crossroads. We hope to investigate this phenomenon further, however.

Finally, consider the points in Figure C.1 with three to ten AFVs in the footprint. The PEM relation fits these points extremely well, though not perfectly. For example, it is disconcerting that more AFVs are killed per salvo when there are five AFVs in the footprint than when there are six or seven.

This is systematic and not due simply to random variation. Each point in Figure C.1 is the average of many tens or hundreds of salvos. While the random variation in kills per individual salvo is large, the random variation on the average of a hundred salvos is much lower.

One possible explanation is that in the DSB '98 simulations, different salvos targeted different mixes of the four types of AFVs (RADAXX, RAPC1F, RAPCXF, and RTANKF), each of which had its own vulnerability to TACMS/BAT. Moreover, early in the simulations, AFV packets could overlap packets of other kinds. Thus many vehicle types might be in the footprint of a single salvo of TACMS missiles. Table C.1 lists the 13 types of vehicles found in the DSB '98 simulations, along with the number of each type in the simulations, its

Table C.1

Numbers and Kill Fractions of Red Vehicles

Vehicle Type	Number	Kill Fraction	Exposure Odds
RADAT	12	0.057	1.03
RADAXX**	4	0.146	1.72
RAPC1F**	35	0.123	1.34
RAPCXF**	32	0.318	0.97
RHELR	36	0.000	0.10
RLCVA	12	0.059	2.00
RLCVAA	27	0.264	0.28
RLCVM	12	0.010	2.00
RLTATK	7	—	0.00
RMCMV	21	0.000	1.40
RMCVGM	12	0.000	0.01
RTANKF**	33	0.250	1.05
RTRK3	300	0.009	1.08

** AFVs.

likelihood of being in the footprint of a salvo of TACMS missiles, and a measure of its vulnerability to TACMS with BAT.

The kill fraction of a vehicle type is the ratio of the number of kills of that type in all the simulated cases to the actual number of exposures. Since the kill fractions of different types of AFVs ranged from 0.123 to 0.318, the maximum error from using the weighted average kill fraction is about 50 percent. The actual error will be less, however, if all AFV packets have about the same proportions of the high kill fraction vehicles (RAPCXF and RTANKF) versus low (RADAXX and RAPC1F). In fact, most AFV packets contained 70 percent or more vehicles with the higher kill fractions. In only 1 AFV packet out of 13 were fewer than half the vehicles either RAPCXF or RTANKF.

The exposure odds for a vehicle type is the ratio of the actual number of appearances of that vehicle type in footprints of simulated TACMS salvos to its pro rata share of the total number of appearances of all vehicles. While all types of AFVs received their pro rata share of exposures or somewhat more, other types of vehicles were equally likely to be exposed, and they often appeared in footprints that also contained AFVs. This was especially likely early in the scenario, before the three different kinds of packets (see Appendix B) had separated. We speculate that BAT might first be attracted into the neighborhood by the noise of AFVs, but the submunitions could be "distracted" by non-AFV vehicles during their IR detection stage.

Barnett, Jeffery R. (1996), *Future War: An Assessment of Aerospace Campaigns in 2010*, Maxwell AFB, AL: Air University Press.

Belldina, Jeremy S., Henry A. Neimeier, Karen W. Pullen, and Richard C. Tepel (1997), "An Application of the Dynamic C4ISR Analytic Performance Evaluation (CAPE) Model," McLean, VA: MITRE Technical Report.

Bingham, Price (1997), *The Battle of Al Khafji and the Future of Surveillance Strike*, Arlington, VA: Aerospace Education Foundation.

Bowie, Christopher, Fred Frostic, Kevin Lewis, John Lund, David Ochmanek, and Philip Propper (1993), *The New Calculus: Analyzing Airpower's Changing Role in Joint Theater Campaigns*, Santa Monica, CA: RAND.

Davis, Paul K., and James Bigelow (1998), *Experiments in Multiresolution Modeling*, Santa Monica, CA: RAND, MR-1004-DARPA.

Davis, Paul K., James Bigelow, and Jimmie McEver (1999), *Analytical Methods for Studies and Experiments on "Transforming the Force,"* Santa Monica, CA: RAND, DB-278-OSD.

Davis, Paul K., James Bigelow, and Jimmie McEver (forthcoming), *Estimating Aggregate Effects of Terrain, Maneuver Tactics, and C4ISR on the Effectiveness of Long Range Precision Fires: the PEM Model*, Santa Monica, CA: RAND.

Davis, Paul K., and Manuel Carrillo (1997), *Exploratory Analysis of the Halt Problem: A Briefing on Methods and Initial Insights*, Santa Monica, CA: RAND, DB-232-OSD.

Davis, Paul K., David Gompert, Richard Hillestad, and Stuart Johnson (1998), *Transforming the Force: Suggestions for DoD Strategy*, Santa Monica, CA: RAND, IP-179.

Davis, Paul K., and Richard Hillestad, *Exploratory Analysis Policy Problems with Massive Uncertainty*, Santa Monica, CA: RAND, forthcoming.

Davis, Paul K., Richard Hillestad, and Natalie Crawford (1997), "Capabilities for Major Regional Conflicts," in Zalmay Khalilzad and David Ochmanek (eds.), *Strategic Appraisal 1997: Strategy and Defense Planning for the 21st Century*, Santa Monica, CA: RAND.

Davis, Paul K., William Schwabe, Bruce Narduli, and Richard Nordin, *Mitigating Effects of Access Problems in Persian Gulf Contingencies*, Santa Monica, CA: RAND, forthcoming (a draft version available within the government was issued in 1997).

Defense Science Board (1998a), *1998 Summer Study Task Force on Joint Operations Superiority in the 21st Century: Integrating Capabilities Underwriting Joint Vision 2010 and Beyond*, Volume 1, Office of the Under Secretary of Defense for Acquisition, Washington, D.C.

Defense Science Board (1998b), *1998 Summer Study Task Force on Joint Operations Superiority in the 21st Century: Integrating Capabilities Underwriting Joint Vision 2010 and Beyond*, Volume 2, supporting analysis, Office of the Under Secretary of Defense for Acquisition, Washington, D.C.

Defense Science Board (1996), *Tactics and Technology for 21st Century Military Superiority*, Volume 2, Office of the Under Secretary of Defense for Acquisition, Washington, D.C.

Helmbold, Robert L. (1990), *Rates of Advance in Historical RAND Combat Operations*, research paper CAA-RP-90-1, U.S. Army Concepts Analysis Agency, Bethesda, MD.

Johnson, Stuart, and Martin Libicki (1995), *Dominant Battlefield Knowledge: The Winning Edge,* Washington, D.C.: National Defense University (reprinted in 1996).

Joint Chiefs of Staff (1997), *Concept for Future Joint Operations: Expanding Joint Vision 2010, Joint Warfighting Center,* Fort Monroe, VA.

Lumina Decision Systems (1996), *Analytica User Guide,* Los Gatos, CA.

Matsumura, John, Randy Steeb, Tom Herbert, M. Lees, Scot Eisenhard, and A. Stich (1997*), Analytic Support to the Defense Science Board: Tactics and Technology for 21st Century Military Superiority,* Santa Monica, CA: RAND, DB-198-A.

Matsumura, John, Randy Steeb, Ernst Isensee, Tom Herbert, Scot Eisenhard, and John Gordon (1999), *Joint Operations Superiority in the 21st Century: Analytic Support to the 1998 Defense Science Board,* Santa Monica, CA: RAND, DB-260-A/OSD.

McEver, Jimmie, Paul K. Davis, and James Bigelow (forthcoming), *EXHALT: An Interdiction Model for Assessing Halt Capabilities in a Larger Scenario Space,* Santa Monica, CA: RAND.

National Research Council (1997), *Modeling and Simulation, Volume 9 on Technology for the United States Navy and Marine Corps, 2000–2035,* Washington, DC: National Academy Press.

Naval Studies Board (1997), *Technology for the United States Navy and Marine Corps, 2000-2035,* National Research Council, Washington, D.C.

Ochmanek, David, Edward Harshberger, David Thaler, and Glenn Kent (1998), *To Find and Not To Yield: How Advances in Information and Firepower Can Transform Theater Warfare,* Santa Monica, CA: RAND.

Ochmanek, David, Glenn A. Kent, Alex Hou, Ernst Isensee, Robert E. Mullins, and Carl Rhodes (unpublished), research on interim results of joint interdiction capabilities assessment, Santa Monica, CA: RAND.

Pace, Dale (1998), "Verification, Validation, and Accreditation," in David J. Cloud and Larry B. Rainey (eds.), *Applied Modeling and Simulation: An Integrated Approach to Development and Operation*, New York: McGraw Hill.

Sovereign, Michael (1995), "DBK with Autonomous Weapons," in Stuart Johnson and Martin Libicki (eds.) (1995), *Dominant Battlefield Knowledge: The Winning Edge*, Washington, D.C.: Institute for National Strategic Studies, National Defense University, 103–113.